CREATIVE **MINDS** CHARMED LIVES

Interviews at
Institute for Mathematical Sciences
National University of Singapore

Yu Kiang Leong
National University of Singapore

CREATIVE MINDS CHARMED LIVES

Interviews at
Institute for Mathematical Sciences
National University of Singapore

World Scientific
NEW JERSEY • LONDON • SINGAPORE • BEIJING • SHANGHAI • HONG KONG • TAIPEI • CHENNAI

Published by

World Scientific Publishing Co. Pte. Ltd.
5 Toh Tuck Link, Singapore 596224
USA office: 27 Warren Street, Suite 401-402, Hackensack, NJ 07601
UK office: 57 Shelton Street, Covent Garden, London WC2H 9HE

British Library Cataloguing-in-Publication Data
A catalogue record for this book is available from the British Library.

CREATIVE MINDS, CHARMED LIVES
Interviews at Institute for Mathematical Sciences, National University of Singapore

ISBN-13 978-981-4317-58-0
ISBN-10 981-4317-58-6

To Professor Jean-Pierre Serre

who kindly gave us the mother of all our interviews

Contents

Contents

Foreword

This interview volume is a collection of all the interviews of prominent visitors to the Institute for Mathematical Sciences, conducted by Yu Kiang Leong and published in the Institute's newsletter *Imprints* from 2003 to 2009.

The concept of a newsletter with interviews as a regular feature was conceived in conversations with Yu Kiang in 2002, two years after the establishment of the Institute. We wanted to publish a newsletter which would be different from the newsletters of other mathematical sciences institutes. We also thought that interviews of successful mathematicians and scientists would be interesting to the general reader and, in particular, inspirational to students. Yu Kiang has a flair for writing, widely acknowledged by all who know him. In 1985, he had also conducted, jointly with Chi Tat Chong, an interview of the legendary French mathematician Jean-Pierre Serre, who became the youngest recipient of the Fields Medal (the mathematical counterpart of the Nobel Prize) in 1954 at the age of 28. The interview was first published in the June 1985 issue of *Mathematical Medley*, a publication of the Singapore Mathematical Society, and later in Volume 8 (1986) of the *Mathematical Intelligencer*. Since then, it has been translated into several languages including Czech, Chinese, Arabic, Japanese and Russian. Since Yu Kiang was available at that time, I invited him to be the founding editor of the newsletter entrusted with the responsibility of interviewing visiting scientists at IMS. He gladly accepted my invitation and the inaugural issue of *Imprints* was published in May 2003 featuring an interview of Roe Goodman. Since then, he has continued to conduct interviews for every issue of *Imprints* although he relinquished his editorship in 2006.

After a number of issues, Yu Kiang and I talked about compiling the interviews and publishing them in a volume at some point. Now that we have a total of 37 interviews and the Institute is approaching its 10th anniversary, it is an appropriate time for us to embark on this endeavor in commemorating the 10th anniversary. I would like to thank Yu Kiang for his invaluable contributions to the conceptualization of the newsletter as founding editor and for his tireless efforts and unfailing enthusiasm in conducting the interviews. I would also like to thank World Scientific for publishing the volume and for its friendly cooperation and help in the design and editing. Thanks also go to the current *Imprints* editors, Denny Leung and Ser Peow Tan, and former editor, Ka Hin Leung, for their careful editing of the newsletter and for ensuring its high quality.

I would like to take this opportunity to express my deep gratitude to all the former and current members of the Scientific Advisory Board — former members Hans Föllmer, Avner Friedman, the late Jacques-Louis Lions and Keith Moffatt; and the current members Roger Howe, Pao Chuen Lui, David Mumford, Olivier Pironneau, David Siegmund and Yum-Tong Siu — for guiding and nurturing the Institute to what it is today, many of them having advised the Institute from its inception 10 years ago.

Louis Chen
Director
Institute for Mathematical Sciences

Preface

Being inexhaustible, life and nature are a constant stimulus for a creative mind.
– Hans Hofmann (1880–1966)
German-born American abstract expressionist painter

After the setting up of the Institute for Mathematical Sciences (IMS) of the National University of Singapore (NUS), Louis Chen, who became its founding director, informally discussed with me his plans to publish a newsletter which would regularly feature interviews of prominent visitors to the Institute. He suggested that I take on the job of being its editor and of conducting and publishing such interviews. Though Chi Tat Chong and I had done a rather well-received interview of the well-known French mathematician Jean-Pierre Serre when he visited NUS in the 1980s, it was only a one-shot thing. Naturally, I was a bit hesitant about doing interviews of distinguished people on such a regular basis. But Louis' optimism, energy and vision persuaded me to give it a try. The first interview was published in 2003 in the first issue of the newsletter the name of which, *Imprints*, was chosen after we had considered many options.

There was some initial apprehension on my part about mathematical interviews. First, mathematicians are generally not known to be enthusiastic about being interviewed, whether it is by nature of their personality or by habit of rigor and economy in thought. Most of them would rather talk mathematics than answer trivial or general questions. Second, how does a non-expert ask meaningful and interesting questions about a deep and specialized subject or at least avoid asking plainly silly questions? Even after these years of interviewing, this feeling still remains.

Being an outsider to a field obviously has disadvantages, but that could also be an advantage. The non-expert interviewer could then be excused for asking questions that seek to elucidate and inform about the relevant field of specialization. After all, readers of a newsletter could come from a wide spectrum of disciplines. Research papers very rarely give any inkling as to how an important idea was conceived; the crucial idea is almost always presented as if it appeared out of the blue in its final form to the discoverer. If he or she could be persuaded during an interview to shed some light on the process that led to the discovery of the idea, it would have been worth the interview. Not many mathematicians or scientists have the time or inclination to write about themselves rather than their research. Perhaps the interview could give us a glimpse of the humanity behind the seemingly emotionally detached endeavor.

While Nobel laureates are expected to be thrown into the interview circuit and make pronouncements on almost anything, Fields Medalists (mathematical counterparts of Nobel laureates) are rarely asked for their opinions on the state of the world, rarely even on the current state of mathematics. And they would not care either. The famous German playwright Johann Wolfgang von Goethe (1749–1832) was reputed to have said, "Mathematicians are like Frenchman: whatever you say to them they translate into their own language, and forthwith it is something entirely different." Unfortunately, the converse is not true. If a mathematician says something to a journalist, the latter may not be able to translate it into a familiar language. Instead, the journalist is often left to despair of cryptic and mathematical (or even metamathematical) remarks. I was, however, more fortunate in that the distinguished visitors I interviewed were generally more than willing to suffer fools, especially in vetting the final versions of the edited transcripts for inaccuracies. But I should hasten to add that any remaining inaccuracy or shortcoming in the final versions would be my sole responsibility.

The visitors to be interviewed were suggested mostly by Louis himself though a few were suggested by program chairs. The lack of well-defined guidelines in the selection resulted in a good cross-section of productive researchers across many areas in logic, Lie groups, number theory, harmonic analysis, combinatorics, partial differential equations, dynamical systems, fluid dynamics, turbulence, probability, stochastic processes, statistics, imaging science, scientific computation, optimization, economics, epidemiology, computational molecular biology and bioinformatics. Perhaps the only guideline was that we should interview every member of the Institute's Scientific Advisory Board.

An interview would involve three stages: preparatory reading of the visitor's background and his or her main research work, the actual interview itself for which a set of proposed questions was sent to him or her beforehand and in which related questions arising from the responses were also asked, and finally, the preparation of an edited transcript which will then be sent for vetting and, in some cases, enhancement. Each of these stages is in itself a multidisciplinary journey of ideas speeding through chronological time and across different countries. It is an educational and exhilarating experience, sometimes dizzying but never dull. To one who is brought up in the tradition of specialization, it is, to use the well-worn cliché, an eye-opening and mind-expanding experience.

The interviews gave me the opportunity to observe the thought processes of creative minds at close quarters and to catch a glimpse of scientific research from a personal view that is in color (even though it may not be in high definition) and in four dimensions (in space and time). One could almost feel the palpable excitement that must have been generated at the high points of scientific discovery and epiphany. Above all, one senses a kind of passion in a search for the Holy Grail, a passion that appears to be both obsessive and compulsive. To the public, the end products of scientific research are objective and devoid of emotion but it is clear that the paths that led to them are often personal and subjective and, at times, fraught with vicissitudes.

It used to be that mathematicians form a closed circle, oblivious and even disdainful of the uses of their esoteric art to practical problems. Occasionally, they would concede that some mathematical formulations of scientific issues could give rise to some interesting questions of purely mathematical interest. On the other side, the engineers and physical scientists are quite happy to cook up their own theoretical recipes and set up *ad hoc* structures to resolve immediate issues without invoking the spell of the mathematicians. But all these have changed. The success of mathematics in general relativity and quantum physics in the first half of the 20th century was followed by the statistical contributions to the Promethean efforts of the Human Genome Project that produced a rough draft in 2000 and an essentially complete version in 2003. Nowadays, the "unreasonable effectiveness of mathematics" (in the words of the physicist Eugene Wigner) is seen to be essential, if only as a necessary evil, to the resolution of specific problems in biology, physics, chemistry, engineering, medical science and economics.

When it was mooted that IMS should celebrate its 10th anniversary this year (2010), Louis suggested that the interviews that appeared in *Imprints* be reprinted as a commemoration volume for the celebration in June 2010. One would naturally ask whether it is worth publishing them other than as a memento for the occasion. No doubt, many things would have changed years after an interview. So would an interview conducted years ago be of interest or relevance to anyone now? It is a matter of course that new questions and newly emerging research areas will now take center stage and old players giving way to fresh faces. But a current scientific edifice often builds on the foundations laid by previous generations. It is more so in mathematics that often there is still something to be mined from an old and well-worked quarry of ideas conceived many, many years ago. Thus interviews can provide an otherwise lost source of history of ideas which will not only interest the historians but which may, in a serendipitous way, illuminate some dark corner of the labyrinths of research.

As we satisfy our thirst and needs from the waters of the many streams that flow from the vast river of knowledge, we hope that this commemoration volume will make us pause and reflect on the springhead of this fast-flowing river of knowledge. In the interviews, we attempt to trace the source in the originators of the ideas and the active practitioners of the science and art, who would in turn trace the source even further back.

The interviews in this volume are essentially reprinted from the past issues of *Imprints* with some minor corrections, changes and updates. On behalf of IMS, I would like to express my deep appreciation to all the distinguished mathematicians and mathematical scientists for the honor of the interviews accorded to the Institute, especially to:

Béla Bollobás, Leonid Bunimovich, Tony Fan-Cheong Chan, Sun-Yung Alice Chang, Jennifer Tour Chayes, Carl de Boor, Persi Diaconis, David Donoho, Robert Engle, Hans Föllmer, Avner Friedman, Roe Goodman, Bryan Grenfell, Takeyuki Hida, Roger Howe, Wilfrid Kendall, Lawrence Klein, Brian Launder, Fanghua Lin, Pao Chuen Lui, Eric Maskin, Eduardo Massad, Daniel McFadden, Keith Moffatt, Stanley Osher, Doug Roble, Ron Shamir, Albert Nikolaevich Shiryaev, David O. Siegmund, Theodore Slaman, Terry Speed, Charles Stein, Gilbert Strang, Eitan Tadmor, Michael Todd, Sergio Verdú, Michael S. Waterman and W. Hugh Woodin.

And personally, I would like to thank all of them for their time, patience and help in this project. It was a great experience for me.

Yu Kiang Leong

Acknowledgements

The author would like to express his sincere thanks and deep appreciation to

(a) all the distinguished mathematicians and mathematical scientists who have kindly granted interviews to the Institute's newsletter *Imprints*, thereby sharing their wisdom and experiences with the scientific community,

(b) the organizers of the various thematic programs and workshops for their suggestions and support in arranging some of the interviews,

(c) all the present and past administrative and technical staff of the Institute for Mathematical Sciences for facilitating the arrangement, conducting and publication of the interviews; especially Claire Tan, Agnes Wu, Stephen Auyong, San Yee Yeoh, Aung Naing Sunn, Cindy Tok, Emily Chan, Wendy Tan and Kian Peng Chua,

(d) all the present and past editors of *Imprints* for their patience, forbearance and suggestions during the preparation of the transcripts of the interviews, namely, Ser Peow Tan, Ka Hin Leung and Denny H. Leung,

(e) Stella Pang and Lynette Wong of the Department of Mathematics, National University of Singapore for their administrative support for the interviews,

(f) Louis Chen for initiating many of the interviews and facilitating communication with some of our distinguished visitors,

(g) Yeneng Sun for his help when he was Deputy Director of the Institute during the formative years of the Institute,

(h) Roberto S. Mariano, Dean of the School of Economics and Social Sciences of Singapore Management University for his help in arranging the interviews of Lawrence Klein and Robert F. Engle,

(i) Kim Song Tan of the School of Economics and Social Sciences of Singapore Management University for co-interviewing Lawrence Klein and Robert F. Engle,

(j) Helen Tombropoulos of the Department of Statistics of Stanford University for her efforts in preparing the final transcript of the interview of David Donoho,

(k) Lynn Friedman for her help in preparing the final transcript of the interview of Avner Friedman,

(l) Kok Khoo Phua for giving the full support of World Scientific to the production of this commemoration volume,

(m) Lai Fun Kwong and her production team (in particular, Bee Keow Ho) at World Scientific for their patience and professionalism given to us during the preparation of this volume as well as during the past years of publication of the interviews in *Imprints*,

(n) all those who have contributed in one way or another to the planning and production of this book.

CREATIVE **MINDS** CHARMED LIVES

Interviews at
Institute for Mathematical Sciences
National University of Singapore

Béla

Graphs Extremal and Random

As long as a branch of science offers an abundance of problems, so long is it alive; a lack of problems foreshadows extinction or the cessation of independent development.

– David Hilbert (1862–1943)
International Congress of Mathematicians, Paris, 1900

Béla Bollobás is well-known for a wide range of significant contributions to graph theory, combinatorics and functional analysis. His recent work on applications of random graph techniques to percolation theory is a ground-breaking contribution to the theoretical basis of a newly emerging field motivated by physical phenomena and first explored by computer simulation.

He may be regarded as a leading exponent of the Hungarian school of graph theory, having paved the way for the current widespread applications of random graphs in numerous areas in applied mathematics, physics and engineering. In addition to more than 350 research papers, he has written 10 books and edited nine volumes. He is also well-known for his mathematical exposition and for championing the cause of the combinatorial approach in mathematics. His two books *Extremal Graph Theory and Random Graphs*, published in 1978 and 1985 respectively, were the first books to systematically present coherent theories of early results in those areas. His latest book *Percolation* is written jointly with Oliver Riordan. Bollobás' personal and mathematical

Bollobás

connections with his mentor, the prolific and consummate problem-solver Paul Erdös (1913–1996) and with his intellectual mainspring Trinity College in Cambridge are the stuff of legends of contemporary mathematics.

A Fellow of Trinity College since 1970, Bollobás has a long and distinguished career at the Department of Pure Mathematics and Statistics in Cambridge University from 1971 to 1996; from 1982 to 1994 he paid long visits to Louisiana State University at Baton Rouge. In 1996, he accepted the Jabie Hardin Chair of Excellence in Combinatorics at the Department of Mathematics of the University of Memphis, Tennessee, while keeping his Fellowship at Trinity College. Since 2005, he has been a Senior Research Fellow of Trinity College. He is also a foreign member of the Hungarian Academy of Science. He has held visiting appointments in various countries throughout the world and has been invited to give lectures at major conferences and scientific meetings. He has supervised over 40 PhD students, some of whom have gone on to distinguished careers, notably Tim Gowers, 1998 Fields Medalist and Rouse Ball Professor of Mathematics at Cambridge University. Bollobás excelled not only in mathematics but also in sports: he represented Oxford University in the pentathlon, and Cambridge University in fencing.

Bollobás' connections with NUS date back to 1994 when he was visiting professor from June to August. During his second visit from May–June 2006 for the Institute's program *Random Graphs and Large-Scale Real-World Networks,* of which he was chair, he was interviewed on 17 May 2006. The following is an edited and enhanced version of the transcript of the interview, in which he traces his mathematical journey from a closed Hungarian communist system to an eclectic academic environment in Cambridge and speaks passionately about his personal mission in spreading the philosophy of combinatorics within mathematics, his reminiscences giving us glimpses of the richness of modern mathematical traditions.

Imprints: You did your first doctorate in Hungary. Who was your supervisor then?

Béla Bollobás: I should be able to answer this question very easily, but I cannot, since in the Hungary of the 1960s we didn't have well-defined supervisors: we would join a group of mathematicians, attend the right seminars, talk to the right people, and work on our dissertations on problems we picked up. The group I joined was that of László Fejes Tóth, who worked on discrete geometry and had written the famous book on the subject, so I wrote my dissertation on packings, coverings, and tilings. However, my real supervisor was Erdös. I had got to know him when I was 14 or so, and from then on he gave me lots of mathematical problems; over the years he kindly stayed in touch with me and inspired me. Of course, he was not in Hungary all that much, but even when he wasn't there, he wrote letters with problems and so he was my real supervisor from the very beginning.

I: Was your education a typically traditional Hungarian one?

B: Yes. I always went to school, didn't stay at home and was not home-schooled like Erdös, for example. But I did have lots and lots of private tutors, not for school work, but for extra-curricular activities. I grew up 5–10 years after the war, and in the communist Hungary of the day many people who had played prominent roles before the war lost not only their livelihoods but even their homes: they were sent into "exile". So I was taught at home by a former general, a count, a baroness and a former judge. They were excellent people, but in those days, they were deemed to be nobodies. So with my education I was exceptionally lucky: I couldn't have had better people to tutor me. The judge was not allowed to remain in the judiciary, so he took up teaching. The general was pretty famous — he was the head of Hungarian fencing. Fencing was actually popular in Hungary for many years. This great-uncle of mine had the wonderful idea of setting up a Fencing Academy in the army, so that able recruits have a chance of being trained to be coaches, rather than go on mindless drills. Within a few years Hungary produced more coaches than the rest of the world put together. Before the war in Central Europe fencing was much more important than it is now: for doctors, judges, lawyers and civil servants fencing played a role somewhat similar to that of golf today. The three countries that were great in fencing were France, Italy and Hungary.

I: Were you spotted by Erdös?

B: In some sense, yes. In Hungary there are many competitions; in fact, the idea of having mathematical competitions at all was born in Hungary. When I was 14, I won the national competition, and, as luck would have it, Erdös had just returned to Hungary for a week or so: he sent word to me that I should go and meet him. I met him in a fancy hotel in Budapest, on a hill-top. We had lunch and it was amazing that he was willing to talk to a 14-year-old boy. He was 45 but to me he looked ancient. Throughout his life he was extremely good to youngsters. His favorite was Louis Pósa, whom he got to know when Pósa was 10 or 11. Erdös was very disappointed when, after a good start to his career, Pósa didn't continue in mathematical research but chose to nurture very talented teenagers.

I: It seems that Hungary has produced a disproportionately large number of mathematicians.

B: That is certainly true. I'm pretty sure it's due to two things. Firstly, in Hungary we had a journal for secondary school pupils. It's a monthly based on attractive and challenging problems. Readers are invited to send in their solutions which are then checked, marked, and the best of which get published. That made a huge difference. The other reason is that there are annual mathematical competitions: three-hour long exams testing your ingenuity on a handful of problems. I believe that the existence of the journal was even more important than the annual competitions, since the competitions in the journal went on throughout the year. All the time, you have problems that you wanted to solve — elegantly. The judges gave you bonus marks if you gave several solutions or you generalized a problem or you sharpened the bounds, which generated much research. Practically everybody I can think of went through this system — Marcel Riesz, Alfréd Haar, Eugene Wigner, von Neumann, Pólya, Szegö, von Kármán. But Erdös was never good at those competitions; von Neumann and von Kármán were very good at them. Wigner and von Neumann were in the same school, and Wigner considered von Neumann to be the only genius he had ever met, although he had known Einstein as well.

I: What made you go to Cambridge to do a second doctorate after your first one in Hungary?

B: Hungary was a very closed-in country. You were not allowed to travel outside, and going abroad was always a tremendous feat. From an early age, I felt claustrophobic. At the beginning of my university studies, I asked Erdös whether I could go and study abroad. I knew that he was allowed to live abroad and came back to Hungary for only short periods. He spent a lot of time in Israel and even had a job there. I asked him whether I could go to Israel for a semester or even a year to study mathematics. Then he said, "Why Israel? You are not even Jewish. Why not Cambridge? I have a very good friend who had just gone there to work with Davenport and maybe he can help you." Of course, going to Cambridge was beyond my wildest dreams. So Erdös wrote a good recommendation to Harold Davenport to try to get me into Cambridge. By then I had a joint paper with Erdös which I wrote when I was still at high school. But we needed permission from the communist authorities. That took ages and ages, and was very humiliating, but eventually I did get the permission and was allowed to go to Cambridge for a year. That was in the middle of my undergraduate studies. After a year in Cambridge I returned to Hungary but very soon I had a scholarship to go Cambridge to do a PhD. I applied to the authorities for permission to go there, but I was refused. Next, I had a scholarship to Paris but was again refused permission to leave the country.

I: You went to Moscow?

B: Yes. After I had my degree, I spent a year in Moscow to work with Israel Moiseievich Gelfand. My year there was a wonderful mathematical experience. After Moscow, but not quite immediately, the communist authorities allowed me to go to Oxford on a scholarship (from Oxford, of course, not Hungary). By then, I said to myself, "If I ever manage to leave Hungary, I won't return." So when I arrived in Oxford, I decided to take up my old scholarship to Cambridge rather than return to Hungary. That way I didn't have to apply for anything because it had been sitting there for years. But then within a year, I got a fellowship from Trinity College, which was better than getting a PhD. There was no pressure on me whatsoever to submit for another PhD. But I thought that as the college had given me a scholarship to do a PhD, it was my duty to get one.

I: I notice that your PhD in Cambridge was done with Adams, who was a topologist.

B: Yes, Adams was my official supervisor but in reality I worked by myself, getting my problems from the Functional Analysis Seminar. When I was in Moscow, Gelfand said that it would be very good to work with Michael Atiyah or Frank Adams, the great topologists. However, when I was in Oxford, Atiyah was on his way to the Institute for Advanced Study, and when I arrived in Cambridge, Adams was still in Manchester although on his way back to Cambridge. By the time Adams arrived a year later, I already had a fellowship at Trinity College. Nevertheless, Adams remained my official supervisor; in fact, I learned a fair amount of algebraic topology from him and I did work on some of his questions. During my first year in Cambridge I joined the functional analysis seminar, where I found several beautiful problems, some of which I solved, so my Cambridge PhD thesis was on Banach algebras.

I: Was your interest in graph theory shaped by your early years in Hungary?

B: It was certainly due to Erdös. If he hadn't been there to give me lots of attractive problems, I'm sure I would have ended up doing either number theory with Turán or probability theory with Rényi.

I: You have published several books on graph theory, including *Extremal Graph Theory* in 1978 and *Random Graphs* in 1985. What made you write these books?

B: It really goes back to the picture I had of graph theory, not only picture but reality. For some peculiar reason, in the early 1970s or later, graph theory came in two flavors; one was done in Western Europe and America, and the other in the East, mostly Hungary by Paul Erdös, Tibor Gallai, Gabriel Dirac and others. In the West, they didn't do any extremal graph theory. On the other hand, in Hungary, graph theory was almost exclusively extremal graph theory. I very much wanted to show that extremal graph theory was a pretty serious subject and not only a collection of random problems that Erdös thought up and popularized. The usual charge against graph theory is, "Ah, it is made up of *ad hoc* problems that have nothing to do with each other. What's the point?" To some extent, at the very beginning, this is true, but slowly, slowly all these results do gel into a single theory, so my aim was to show that there is such a theory — extremal graph theory. I started to write this book very soon after I arrived in Cambridge but it took me ages to finish. I had to take a sabbatical to find enough time to finish it.

The theory of random graphs was founded by Erdös and his good friend Alfréd Rényi in the late 1950s and early 1960s. At the beginning, they wrote several joint papers but the whole theory didn't take off. People didn't jump on it and said, "How exciting! Let's try to continue it." The climate started to change in the 1970s. In particular, Erdös came to visit me in Cambridge for a term as a Visiting Fellow Commoner in Trinity College — perhaps the longest period he spent in one place for many decades, since he never stayed anywhere for more than a week or so. He suggested that we work on random graph problems. I got interested and from then on, I was doing random graphs. I had this urge to showcase the classical theory together with lots of new developments and show that it is not only a beautiful subject but also very important. Really, random graphs became more and more active in those days. Once you write a book, parts of it became outdated almost immediately. It was the first serious book on random graphs just as the book on extremal graph theory was the first book on the subject. They are on different aspects of graph theory but they are closely connected.

I: How do you see the future of combinatorics, especially random graphs?

B: Hilbert, I think, said that a subject is alive only if it has an abundance of problems. It is exactly this that makes combinatorics very much alive. I have no doubt that combinatorics will be around in a hundred years from now. It will be a completely different subject but it will still flourish simply because it still has many, many problems. The same applies to random graphs. In fact, the field of random graphs has connections with statistical physics, percolation theory and even computer science. It's very strange that just at about the same time that random graphs were founded, Broadbent and Hammersley founded percolation theory. These two subjects are all about random subgraphs of certain graphs. They should be about the same — okay, one is finite and the other is mostly infinite and lattice-like, but still, they have about the same questions. For many, many years there were no interactions between the two subjects, none whatsoever. Now this is changing quite a bit. Quite a few combinatorialists are doing percolation-type problems.

I: Are random graphs applied to biology?

B: Yes. In the last 10 years or so, many new spaces of random graphs have been defined in the hope of modeling phenomena in various areas, including biology. People have realized that large-scale real-world networks resemble random graphs. You can't really say that they have this structure or that structure. But this random graph is very different from the classical Erdös–Rényi model of a random graph. It has different characteristics: for example, the degree distribution may follow a power law, unlike in the classical case. One of the main advocates of using new models of random graphs is László Barabási, who also proposed several interesting models.

I: Was the power law discovered empirically?

B: Yes, it was observed that several graphs seem to obey a power law, but there were no proofs that they really do. Physicists and experimentalists have a very different attitude from that of mathematicians: much of the time they are not very interested in rigorous proofs. For a mathematician, it is rather annoying that proving even the basic results about these new models can be pretty tough. Oliver Riordan and I have done a fair amount of rigorous work on properties of power law graphs.

I: You mentioned that there is an abundance of problems in combinatorics. It seems that combinatorial problems are very easy to formulate but very hard to solve.

B: For me, the difference between combinatorics and the rest of mathematics is that in combinatorics we are terribly keen to solve one particular problem by whatever means we can find. So if you can point us in the direction of a tool that may be used to attack a problem, we shall be delighted and grateful, and we'll try to use your tool. However, if there are no tools in sight then we don't give up but we'll try to use whatever we have access to: bare hands, ingenuity, and even the kitchen sink. Nevertheless, it is a big mistake to believe that in combinatorics we are against using tools — not at all. We much prefer to get help from "mainstream" mathematics rather than use "combinatorial" methods only, but this help is rarely forthcoming. However, I am happy to say that the landscape is changing.

When Erdös and Rényi started the theory of random graphs, they had to make do with basic probabilistic results concerning sieves and moments, but combinatorics changed the landscape of probability theory considerably. In order to answer questions in probabilistic combinatorics, results of a different flavor had to be proved in probability theory: results concerning sharp thresholds, isoperimetric inequalities, rapidly mixing random walks, and so on.

There are many other tools as well: algebraic, analytical, and even topological. For example, Borsuk's theorem has been used to prove several beautiful results in combinatorics. The achievement is not in applying such a theorem, after all, every schoolboy knows the theorem, but in discovering that it can be applied, and how it can be applied.

A totally ignorant and unfair way of judging a result in combinatorics is to ask the author: "What have you used to prove your theorem?" Then, upon being told that such and such a theorem was used, comes the retort: "Oh, that's very easy. I could have done it." What nonsense. Yes, of course it's easy once you are told what to do. The achievement is in finding the tool that can crack the problem after a series of clever manipulations that make the problem amenable to the application of the tool.

I: Could the difficulty of combinatorial problems be due to the discreteness of the objects?

B: Not really. Frequently, it is fairly easy to change a discrete problem into a continuous one but more often than not this change does not bring us any closer to a solution. The trouble with the combinatorial problems is that they do not fit into the existing mathematical theories. They are not about functions, topological spaces, groups or operators. More often than not, we simply do not have the machinery to attack our problems. This is certainly not the situation in other branches of mathematics. In fact, it may happen that first a wonderful machine is built and then the search starts for a worthwhile problem that this machine can be applied to. This attitude is totally foreign to combinatorics. In combinatorics we have our problem which at the beginning looks like a Chinese box: there seems to be no way in, there is no indication as to how to start it. Here's the problem: we want to solve it and we don't care in what way we solve it.

I: So you are almost starting from nothing or from the bare minimum...

B: To some extent, yes, but of course, these problems are also built on top of each other. Once a problem gets solved, another one arises, and the theory does build upwards as well, not only sideways. A problem I certainly love and I'm sure is very deep is the problem of conformal invariance in percolation theory. I also love the related problems about the existence of various critical exponents. I have no doubt that these beautiful problems are so hard that they'll be around for many, many years. The original problems are combinatorial although they can also be considered to be problems in analysis or probability theory. I'd be surprised if we didn't need totally novel ideas to solve them.

I: Going the other way, are there any problems in more traditional areas of mathematics that can be solved by combinatorial methods?

B: Oh yes. It is frequently the case that once you have applied all the tools at your disposal, at the end you have to solve an essentially combinatorial problem in the traditional sense: you have to argue from the bits of information you have better than anybody else.

I: I think that the perception of combinatorics has changed considerably.

B: I hope that it is changing, for it should certainly change. Combinatorics is becoming a more "serious" subject, closer to the traditional branches of mathematics — there's no doubt about this. Combinatorics has many really hard questions, like number theory, algebraic topology and algebraic geometry.

I: Is there a single result or discovery of yours that has given you the greatest satisfaction?

B: I wonder how many people can say "yes" to such a question. There are quite a few results that made me very happy at the time, but not one that I would trade for the rest. Let me tell you about some of my favorite results. Not surprisingly, people often like results they proved when they were young.

Thus, I rather like a certain lemma of mine that I proved when I was an undergraduate. It is still one of the very few proper exact extremal results about hypergraphs. (Hypergraphs tend to be nastier than graphs, so this may not be so surprising.) Also, it can be applied in lots and lots of ways. It can be proved very easily: some years after I discovered it, Gyula Katona gave a ridiculously easy and very beautiful proof. But still, I am happy that I found it when I was an undergraduate.

Also, in the early '70s, I wrote a paper with Erdös in which we greatly improved a 30-year-old fundamental result of his, the so-called Erdös–Stone theorem. This theorem says that if a graph G on n vertices has εn^2 more (so, really, very few more) edges than the number guaranteeing a complete subgraph on r vertices, then suddenly it has a complete r-partite graph with t vertices in each class, i.e., r disjoint classes of t vertices, with an edge joining every pair of vertices belonging to different classes. (A little more precisely, we take $r \geq 2$ and $\varepsilon > 0$ fixed, and let $n \to \infty$.) This is very much a "phase transition" type result: once the number of edges increases beyond the point at which a "very thin" complete r-partite graph can be guaranteed, we can guarantee a rather "thick" (t-thick) complete r-partite graph as well. The question is all about the largest t one can guarantee. Erdös and Stone proved that the largest t one can guarantee is at least the $(r-1)^{\text{th}}$ iterated logarithm of n, the order of the graph. Erdös conjectured in numerous papers that the correct bound is precisely this iterated logarithm. To our great surprise, in the early '70s, almost 30 years after the publication of the Erdös–Stone theorem, we proved that the bound is log n, much larger than we imagined.

Another result I do like very much is about the scaling window in the phase transition of a random graph. Let us take a set of n vertices and add to it edges one by one, at random, with the uniform distribution, so that at "time" t we have t edges. The question we are interested in is, "What does this random graph look like at various times?" (Here and elsewhere, all assertions are claimed to hold "with high probability", i.e., with probability tending to 1.) We are mostly interested in one of the crudest properties of our random graph: the number of vertices in the largest connected component. The greatest discovery of Erdös and Rényi was that at time $n/2$ there is a sudden *phase transition* in the sense that if the number of edges is a little less than $n/2$ then there is no large component, in fact, every component has at most order log n vertices; however, if the number of edges is $cn/2$ for some constant $c > 1$, then suddenly there is a *giant component*, a component of order n, in fact, a component with about $\alpha(c)n$ vertices, where $\alpha(c) > 0$. So the size jumps from order log n to order n.

Although at first sight this is a sharp result, it is far from so. Let us look at the point of phase transition through a magnifying lens. What magnification should our lens have to enable us to see the continuous emergence of the giant component? More formally, let us look at our process at time $t = n/2+s$. For what values of s is the largest component much larger than the second? Here are two rather different scenarios consistent with the theorem above. (1) If $s > n/\log n$ then with high probability the maximal component is at least 10^{10} times as large as the second, while for $s < n/(2 \log n)$ this is false. (2) If $s > n^{1/2}$ then with high probability the maximal component is at least 10^{10} times as large as the second, while for $s < n^{1/2}/\log n$ this is false. Now, in the first case we would say that the *window* of the phase transition is about $n/\log n$, while in the second the window is about $n^{1/2}$.

About a quarter century after Erdös and Rényi proved their famous result, I proved that the size of the window is, in fact, $n^{2/3}$. Furthermore, if s is substantially larger than $n^{2/3}$, say, $s \geq n^{2/3} \log n$, but is still $o(n)$, then the largest component has about $4s$ vertices, and all other components are *much* smaller. This was the very first rigorous result about the size of a non-trivial window. All this is, of course, very close to percolation.

Let me finish with two more results. First, a lovely little theorem I proved with Andrew Thomason, which really should have been proved 150 years ago by Steiner or another geometer. Take any d-dimensional body of volume one. In that case, I can give you a box (a rectangular parallelepiped), also of volume one, so that no matter on which plane you project your body and the box, the projection of the box has at most as big a volume as the projection of the body. Note that we are talking about projections into $2^d - 2$ non-trivial subspaces: d subspaces of dimension 1, $d(d-1)/2$ subspaces of dimension 2, and so on. It is a little surprising that there is a body that in this sense minimizes all these projections.

And the last. Very recently, Oliver Riordan (one of the co-organizers of this program) and I proved that the critical probability of random Voronoi percolation in the plane is one-half. Of course, everybody who knows a little about percolation would have sworn that this critical probability must be 1/2 and nothing else, but proving it was a very different matter. There is a strong similarity with the events in the 1960s and 1970s, when everybody in percolation theory knew that the critical probability of bond percolation on the square lattice was 1/2, but nobody could prove it; eventually, after a 10-year gap, Harry Kesten found a proof. The question concerning Voronoi percolation turned out to be much more complicated than that about the square lattice; my paper with Oliver will be published soon. Actually, our hope was that it would be the first step towards proving conformal invariance for random Voronoi percolation. The trouble is that even the "preliminary step" of showing that the critical probability is 1/2 was much more difficult than we had bargained for, so we haven't yet managed to make much progress with conformal invariance.

I: It seems that you are a counter-example to the belief that good results can only be obtained before the age of 40.

B: Maybe, maybe, but, of course, the belief that a mathematician is dead after the age of 40 is very much the figment of G. H. Hardy's imagination. Hardy loved to say that only young man can do real mathematics when, in fact, he himself was a very strong counter-example to that. Hardy after 40 was much, much better than Hardy before 40.

I: You have quite a few research students. Do you like teaching them?

B: I love to have good students. One of the many reasons why I love to be in Cambridge is that Cambridge has by far the best research students in Britain. I have had over 40 research students, many of them extremely good. It would be wrong to list them because whomever I wouldn't mention would be right to feel slighted. But let me just say that four of my students are professors in Cambridge. One of them is a Fields Medalist — Tim Gowers. His is the only name that I consider legal to mention because he's the only one to have got a Fields Medal.

I: Who are the people who influenced you most?

B: Paul Erdös is clearly the man who influenced my mathematical career the most. He was at almost every conference that I attended for 25 years. And one of my jobs at these conferences was to look after him. I really enjoyed his company very much. I would not have imagined how much I would miss him: I am really surprised that even a decade later I miss him very much.

When I was at Cambridge as an undergraduate, I got to know the great physicist Paul Dirac and his wife very well; I became very much part of their family. Mrs Dirac was from Hungary: she was a sister of Eugene Wigner, the Nobel-Prize-winning physicist. It was wonderful to be around the Diracs. Mrs Dirac was the best hostess I have ever seen: she was very well read, had a great appreciation of art, loved antiques of all kind, and was extremely skillful to move the conversation to interesting, unconventional topics. Paul Dirac was an absolutely "free man", the free man *par excellence*, free in the sense that he was free of convention, and didn't have any baggage to carry, as he didn't want to prove himself, and did not mind what people thought about him. He was very polite and considerate, but he could say quietly his own opinion which was often different from that of other people's.

I: He was well-known for not saying too much, wasn't he?

B: That's true, but he did talk quite a lot when he was among friends. He talked to me quite a lot; I could never complain that he didn't. He is someone I have always respected tremendously. Unfortunately, precisely when we moved to Cambridge from Oxford in 1969, he retired to live in Tallahassee, Florida. It was a great blow to us because the Diracs were the people we knew most intimately in Cambridge. From then on, we always went to visit them in Tallahassee and stayed there for a week or even a month. People in Cambridge could never understand what Dirac could be doing in Florida, how he could "put up" with Florida after Cambridge. However, Dirac loved to be in Tallahassee and often told me that he should have moved there much earlier.

When I arrived in Cambridge for good, to become a Fellow of Trinity College, I was surprised that J. E. Littlewood was still alive, as to me he was quite legendary. I was amazed that he was still around in the college. It was mostly through my wife, Gabriella, that I got to know him very well, and I am very lucky that I did. Gabriella, who is a sculptor, made several busts of him; one of these is now in the Combination Room of Trinity College. Littlewood had the reputation of being totally unapproachable, but by the time I got to Cambridge, he had mellowed much. Unfortunately, most of his former students and colleagues still respected him too much and were also a little afraid of him, so they very rarely visited him. He came to have dinner with us a lot; many times. When we had people for dinner, we asked him as well; his presence lent a weight to the evening as everybody was honored to be at dinner with Littlewood.

I: How old was he then?

B: He was 85 when we got to know him, and died at 92. He loved mathematics and had many stories about his friends, including Hardy, Russell and Wittgenstein. While sitting in the Combination Room, sipping claret, he would start his story with "Before the war..." Whenever somebody would ask "Which one?" the answer was always "The first." That was really wonderful.

When he died, I became his literary executive and inherited all his letters and papers; many of these papers originally had come from Hardy. I edited a collection of his stories, *Littlewood's Miscellany*, which is a delightful book, about twice as long as its predecessor [*A Mathematician's Miscellany*] and has many more stories. Of course, the stories were not new, but he remembered them after he had published that book. The extended version was published only after he died.

I: What do you think about Erdös' idea of the "ideal proof from the Book"?

B: Not very much. Actually that was really a joke of his — I talked about this with him many times. He was interested in proving good results; he did not set out to find the proof from the Book, as has been said about him many times. Of course, he was particularly pleased to find beautiful proofs of *simple results*. He always said, "Look, such gems, such really simple, beautiful proofs can only be found in the Book." You don't expect the Riemann Hypothesis to have a proof from the Book that one can give in five minutes. Of course not. You would expect an infinitely more complicated proof. So he always used "The Book" as a joke to enliven his lectures; it should not be taken seriously.

I: You have positions at Memphis and Cambridge. Isn't that a strange combination?

B: I must admit that it is. Everybody thinks it is. Actually I love both places very much. Cambridge is our true home: that's where we have been for close to 40 years, and that is where our real house is — I'm sure that eventually we shall live only in Cambridge, with occasional trips to Budapest. But we also love to be in Memphis.

When I say that I love Memphis, people tend to be puzzled, but they don't know what they are talking about. In the first instance, we went to Memphis because my wife got absolutely fed up with Cambridge, finding it claustrophobic, and Erdös suggested that I go to Memphis, which he had visited many times, often several times a year. In Memphis I have a really wonderful job — no lecturing, no administration, a great assistant to look after me, funds to invite visitors, funds to travel, very clever and kind colleagues, an excellent gym, and so on. Although I do not have to lecture, I always give a graduate course on a topic I hope to write a book on. I view Memphis as a mathematical training camp, where the first thing to do is mathematics, and there is no second. Erdös had very good friends at Memphis — Ralph Faudree, Dick Schelp, Cecil Rousseau, Chip Ordman — mathematicians who helped him a lot: they are still in Memphis and now they are my friends as well; since my arrival they have been joined by several other excellent people like Paul Balister, Vladimir Nikiforov and Jenö Lehel.

On the other hand, when I say that I love Cambridge, nobody is surprised: "Of course, Cambridge is great." And Cambridge is great. I don't know whether you have been to any of the Cambridge colleges.

For me one of the best aspects of my own college, Trinity, is that academics from different disciplines mix: we have outstanding people from all kinds of different subjects at our fairly informal lunches and rather formal dinners. You may find yourself sitting next to a physicist and an economist, and opposite a historian and a physiologist. These are wonderful occasions: you can talk about a great variety of topics to real experts in those fields. Also, it is flattering to be in a place where so many excellent people work. Of course, many a first-time visitor misses this aspect of a college entirely since with him the conversation tends to be shallow: "How long are you staying in Cambridge?", "Have you been here before?", "Where do you come from?", and so on. Thus, Vladimir Arnold got it completely wrong when he imagined that this kind of conversation goes on all the time. This couldn't be further from the truth.

I: Do you have a special position in Memphis? Was it created for you?

B: I'm the first occupant of a rather special chair, the Jabie Hardin Chair in Combinatorics. This chair was not created for me, but Erdös persuaded me that I should accept it, and my colleagues in Memphis were kind enough to be happy about it.

I: Do you travel a lot?

B: Yes, I do — too much. I'm sure the urge to travel goes back to my childhood. In Hungary I grew up feeling imprisoned, and I was always longing to travel, especially to the South. I still find the South very romantic.

I: Erdös traveled a lot too.

B: Yes, Erdös traveled an awful lot. He traveled in a different way, he traveled alone, and almost always went for rather short periods. I frequently go for several months, and then I take lots of people with me, mostly my students and former students from Cambridge and Memphis. I feel that I have to take my current students with me if I want to take care of them: it would be very unfair to leave them at home.

I: I understand that you have taught our present Prime Minister Lee Hsien Loong.

B: I certainly taught him more than anybody else in Cambridge. I can truthfully say that he was an exceptionally good student. I'm not sure that this is really known in Singapore. "Because he's now the Prime Minister," people may say, "Oh, you would say he was good." No, he was truly outstanding: he was head and shoulders above the rest of the students. He was not only the first, but the gap between him and the man who came in second was huge.

I: I believe he did double honors in mathematics and computer science.

B: I think that he did computer science (after mathematics) mostly because his father didn't want him to stay in pure mathematics. Loong was not only hardworking, conscientious and professional, but he was also very inventive. All the signs indicated that he would have been a world-class research mathematician. I'm sure his father never realized how exceptional Loong was. He thought Loong was very good. No, Loong was much better than that. When I tried to tell Lee Kuan Yew, "Look, your son is phenomenally good: you should encourage him to do mathematics," then he implied that that was impossible, since as a top-flight professional mathematician Loong would leave Singapore for Princeton, Harvard or Cambridge, and that would send the wrong signal to the people in Singapore. And I have to agree that this was a very good point indeed. Now I am even more impressed by Lee Hsien Loong than I was all those years ago, and I am very proud that I taught him; he seems to be doing very well. I have come round to thinking that it was indeed good for him to go into politics; he can certainly make an awful lot of difference.

I: Do you have any books in the pipeline?

B: I have two books coming out for the International Congress in August. One of them is a collection of problems — lots of beautiful problems, exactly what we discussed over coffee in Memphis with Paul Balister and others. It will be published by Cambridge University Press and is called *The Art of Mathematics* with the subtitle *Coffee Time in Memphis*. The other one is a book I wrote jointly with Oliver Riordan: its title is just *Percolation* — short and punchy.

Leonid

Stable Islands, Chaotic Seas

Leonid Bunimovich has contributed to the fundamental understanding of dynamical systems and made important applications of probability and statistics to geophysical hydrodynamics, neuroscience, operations research, statistical mechanics, mathematical biology and numerous other scientific areas.

Bred in the great Russian tradition of probability, statistics and mathematical physics of the well-known and influential mathematician Yakov Sinai, Bunimovich began his research interests in Moscow University and quickly developed his own original and independent approaches to various problems in genetics, geophysics, biology, statistical physics and other scientific areas outside mathematics — problems that he encountered in his scientific journey that crossed the high seas and great continents. Even before the political convulsions that shook and broke up the former Soviet Union in 1991 took place, his seemingly chaotic path in academia finally found a niche for him in the School of Mathematics of the Georgia Institute of Technology. Holding the Regents' Professorship, he was the Director of the Southeast Applied Analysis Center at Georgia Institute of Technology, and continues as the director of its successor program — the Applied

Bunimovich

and Biological Contemporary Mathematics Program. He is a winner of the Humboldt Prize and serves on the editorial boards of numerous leading international journals on applied mathematics and mathematical physics. He is actively engaged in organizational work for many scientific meetings around the world. He has traveled widely to major research centers as visiting professor and been invited to speak at major scientific meetings around the world.

Bunimovich was in the organizing committee of the Institute's program on *Dynamical Chaos and Non-Equilibrium Statistical Mechanics: From Rigorous Results to Applications in Nano-Systems* held from 1 August to 30 September 2006. He was interviewed on 17 August 2006. The following is an edited and enhanced version of the transcript of the interview in which he traced the unusual scientific odyssey that took him physically to many places and scientifically into numerous disciplines which speak the language of mathematics. He gives us a first-hand account of scientific discovery and a bird's eye view of the enigmatic landscape at the interface of physics and mathematics which underlies the tantalizing field of chaotic dynamics. He also gives us an insight into the politics of a monolithic ideology that impeded the development of the biological sciences in the vastest political empire that dominated the world in the 20th century.

Imprints: Your PhD in Moscow was in probability and mathematical statistics while your Doctor of Science was in theoretical and mathematical physics. Was there a switch of research interest or was it more of a "natural" transition of research interest? What motivated it?

Leonid Bunimovich: It was not a switch of research interest. I graduated from the department of probability theory and was a student of Professor Sinai who is the major person in dynamical systems and one of the greatest mathematicians of our time. He was a student of Kolmogorov. I was working on the statistical properties of dynamical systems. This again goes back to Kolmogorov. The general view is that there are random phenomena, and there are deterministic phenomena, but in his short paper of 1958, Kolmogorov built a bridge between the world of random systems and the world of deterministic systems. This was the major event that started the revival of ergodic theory, which is the statistical theory of dynamical systems. At that time, it was considered to be essentially a finished area of mathematics. All of a sudden it started to evolve and grow enormously. This is considered to be one of the major developments in science, not only in mathematics, in the 20th century. My thesis was in stochasticity of dynamical systems, where a new mechanism of stochasticity, later called a mechanism of defocusing, was discovered. Ergodic theory was created in the works of Boltzmann and Gibbs on statistical mechanics. My Doctor of Science dissertation was on applications of these new ideas in ergodic theory to statistical mechanics.

After getting my PhD from Moscow University, I didn't work a single day as a mathematician in the Soviet Union because of political reasons. This is why there was such a long gap between my two dissertations as well. I "was allowed" to defend my Doctor of Science dissertation after perestroika started and there was some kind of transition, but scientifically, there were no real changes besides a possibility to travel to scientific meetings abroad. I worked in many areas, but because of the strange, not usual way, I did not work as a mathematician after graduating from what was considered the best university in the world and defending an outstanding PhD. The same thing happened to many young mathematicians of Jewish origin. So I worked in many places. Maybe that is why my scientific interests and the questions I worked on are broad and perhaps even strangely broad.

I: You were actually applying a lot of mathematics to other problems after your PhD.

B: That is exactly what happened. I needed to work somewhere and couldn't find a job, not only me. The political situation was very bad and I believe that the disintegration of the Soviet Union started at that time; it was just concluded in 1991. I had to support my family, I was already married. So I found a job — it was 32nd place where I applied. It was at the Institute of Psychiatry of the Academy of Medical Sciences. They wanted to hire a mathematician because the institute had a computer. Two years later, when the computer broke down, they asked me to repair it. They thought that since it was a mathematical machine, mathematicians must be able to repair it. By then I had done some work in psychiatry and genetics. They realized that I could do something there besides repairing computers. So they hired an engineer to repair it, and I retained and continued my job there.

I: It seems that your scientific journey from Moscow University to Georgia Institute of Technology has been a very long one. Was it a planned one or did it just happen?

B: It was not planned at all. It was life with all its turns and changes, so on and so forth. For instance, my first job was in psychiatry and I applied mathematics to real problems in psychiatry and genetics. It is again the history of the Soviet Union. I was the only mathematician speaking at the First All-Union Conference of Medical Genetics. In Stalin's time, genetics was considered as capitalist science; it was forbidden, and many researchers working in this area were sent to camps. In high school, we never studied genetics. I learned it only after coming to the Institute of Psychiatry and was fascinated by it. There were already new developments. Geneticists who had spent some time in prisons and camps and survived there came back and were working again on genetics, which again became an "allowed" science. They were much older than me. I was then under 30. When there were PhD defenses, it was very interesting for me. There were two people needed to read the thesis — the first was usually very senior, and the second sometimes was me if mathematics was used in the thesis. After the defense, it was the Russian tradition to have a banquet. That was where I learned a lot of real history of Soviet genetics. I wasn't doing mathematics, but I was doing an exciting science, and many people suffered much more than me. I couldn't complain.

I: Is the banquet after the defense organized by the department?

B: It's usually organized by the person who made the defense. It was just a traditional celebration. It is paid for by himself, but more often by his parents. It was the tradition that parents were very happy to have educated kids and sacrificed a lot.

I: Your research interests are wide ranging. Do you think that the Russian system of education has something to do with the range of your research interests and inclination?

B: It depends. In my case, yes. As everybody knows, Kolmogorov was one of the greatest mathematicians of the last century. Take his work on turbulence — it's the basis of turbulence theory for physicists. He was a mathematician, but he laid the foundations of modern turbulence in a four-page paper. This is the style that I always admire. New clear ideas lead to some clear implications for real world problems. The longest time that I worked in the Soviet Union was in the Institute of Oceanology. This is, in fact, my third trip to Singapore. The first time that I came here was as a sailor essentially. Twenty-six years ago, I came with a scientific ship. I came here the second time, two years ago, and the changes in Singapore were very impressive.

I: Did you do any kind of experiments on board the ship?

B: I was a theoretician there, but almost everybody was an experimentalist. It was a long journey, about four months. I was young and strong and could help to carry heavy instruments, not just do theory. Most of the journey was devoted to the experimental studies of the oceans. On the way back, it was another thing. I had only a short time to somehow think over the results and to come up with simple models that would show that the results of measurements were correct and novel. I learned also to value the work of the experimentalists and how to talk to them, although they talked a different "language". Kolmogorov had also been on such a journey earlier on. He was there when his theory was under investigation. He really wanted people to check his theory with experiments. Then he published another paper which took into account the measurements. Of course, it was his influence. It was the style of his school. For example, when I was an undergraduate, my supervisor Professor Sinai would tell me about some dynamical system and said, "This is an interesting system. Look into that." But he did not always tell me what I should prove. By the way, some of the problems Sinai brought up came out of the research of one of the organizers of this program, Professor Zaslavsky, who was a physicist at Novosibirsk at that time.

I: The Russian tradition seems to be that theoreticians, even pure mathematicians, have a close interest in data and experiments.

B: It's actually not like that. A great majority of pure mathematicians, because of the same reasons as mine, did not work as mathematicians but worked in some applied institutions. They naturally didn't like that and were not much interested in what was going on around them. Most of them were really pure mathematicians and didn't really understand what the physicists and engineers were talking about. They don't give definitions. It's a kind of personal thing whether you are willing to understand the things which are not exactly defined. For me, a formulation of a new mathematical problem and proof of that for some natural (hopefully visual as well) examples is much more important and exciting than trying to prove this for more and more and more general classes of systems. More and more technical ideas are needed for that, often combinations of various techniques, but I always prefer simple proofs. Now, sometimes a mathematical community gets embarrassed if a simple proof is found for a long-standing problem, whereas it should be to the contrary.

I: Would you consider yourself to be some kind of theoretical physicist?

B: In fact, some of my friends and colleagues call me a physicist. I don't think there is a big difference. I don't have a broad training and background in physics though it was part of our education. But I'm really interested in physical problems. It also depends on your scientific taste. Many of my results are just examples, and you can build a lot of generalizations on them. I'm more interested in the phenomena, maybe it's a more physical approach. I think all this is science. What I don't like in the

US, for instance, is that they always say "Mathematics and Science". What is really part of the Russian mathematical system is that there are no sharp borders between scientific disciplines. If you remember, Francis Bacon, founder of natural philosophy, said, "Any science reaches a really high level only when it manages to use mathematics."

I: Is chaotic dynamics a recent development of the chaos theory of the seventies?

B: "Chaos" is (actually was for a long time) a very good selling word. Chaos is just a part of that new branch of science which Kolmogorov founded in 1958, 20 years before the word "chaos" was coined. It was called stochasticity of dynamical systems, which means that dynamical systems, purely deterministic systems, can demonstrate the same behavior as purely random systems. It was a real physical and even philosophical discovery. To the general public or people who give funds, what is "stochasticity of dynamical systems"? It is something vague. So "chaos" was coined and chaos is only part of this general area of stochasticity of dynamical systems. Chaotic dynamics is just one face of complex dynamics. The first book on this subject was published by George Zaslavsky in 1970 under the title *Stochasticity of Dynamical Systems*. But "chaos" completely took over. You know, when something becomes more fashionable, you give up something else.

I: The term "chaotic dynamics" seems to suggest more of a physics discipline.

B: It's not only a branch of physics. It's a branch of science — it's also chemistry, biology, geology, geophysics and many other disciplines. But, of course, first of all, physics. Physicists are mathematically trained and they can use the computer better than mathematicians. This is why it was first used in physics. There are many physical systems that develop chaotic behavior. What is the basis of that? Historically, what people knew for centuries, starting with Laplace and even before, is that if you knew exactly the initial conditions of your equations, and the functions involved are sufficiently smooth, then there is a unique solution that can completely predict the evolution of the system. But in practice, if you have any measuring device like a thermometer in physics or medicine, you never have complete precision. You know approximately how you drive your car, approximately 50 mph, maybe 51, but not 51.603. In any practical situation, you work with some such small set of data, not a point. You study the evolution of this small set, and very often it does not look like the evolution of points. This is where all the chaotic dynamics occurs. You have very good precision at the beginning, but with time you lose it. Your prediction can only be statistical.

I: What are some of the central problems and recent advances of chaotic dynamics?

B: This is a kind of natural evolution and development. The major discoveries were in the late '50s and '60s by Kolmogorov, Sinai, Smale, Arnold, Moser, Anosov. Dynamical systems evolution can be very complicated. Another of Kolmogorov's work said that not only a dynamical system's evolution can be complex, but the simplest (integrable) dynamics in Hamiltonian systems is actually stable. It is the celebrated Kolmogorov–Arnold–Moser (KAM-) theory. In practice, you see all those things that are stable under small perturbations. There was a general belief though in physics that if you take a surface (manifold) of constant energy, then the motion is ergodic there, uniformly distributed, but KAM-theory said that it's the opposite situation if you have integrability. Integrability is stable, chaoticity is stable as well. The studies so far took care of these two polar situations — complete chaos and near integrability. The most challenging problem now is: what is in between? The system is neither integrable nor chaotic. Instead, it has a mixed behavior — sometimes it's divided phase space — you have islands of stability in phase space that are called KAM-islands, situated in a chaotic sea. It is much more difficult to study such intermediate systems. Some methods were developed to study the chaotic dynamics, another totally different methods were developed to study stable dynamics. But, at the border of these islands in chaotic seas, you cannot apply any of these methods. This is the major challenge and development.

Dynamical systems behave like stochastic processes. First of all, we are looking at the simplest random processes — coin tossing, independent random variables. But most often, in applications you don't see independent random variables, you don't see Markov processes, but you see processes with infinite but decaying memory. This is a much more difficult problem. There were some breakthroughs. For instance, the simplest examples were found which demonstrate the co-existence of islands and chaotic dynamical systems. By studying these examples, one can completely analyze the system and generalize the theory. The basis of the theory of dynamical systems and ergodic theory often comes from some simple classical examples. One of the major efforts now is the study of the so-called Arnold diffusion — if you start in the chaotic region, how does the particle move between the islands? Does it move fast or slowly, can it move far and so on?

Another development is related to general questions in communications theory and biology, and it has to do with interacting dynamical systems, like systems of neurons, communication networks. We now know rather well how finite-dimensional dynamical systems may behave. But suppose you have several such systems that are connected. Then some new general questions appear. How does the whole system behave? It's space-time dynamics. Not only dynamics in time, but in space because you have different local systems (elements of a network). How do networks behave? It raises questions about different types of synchronization, space-time chaos, etc.

I: Are there are general results for such questions?

B: There are very few results so far; only for some special classes of dynamical networks. But there is no general theory. This is a major challenge.

I: Is there any theory for infinite-dimensional dynamical systems?

B: Again for some classes, such a theory exists, but usually it is not something which is likely to have real applications. But it is very important to find a class (even a narrow one) of systems where we can understand everything. It helps to build intuition on what to expect in the evolution of more general networks.

I: You did some work on chaotic motion of billiards. Is it related to the Hadamard billiards introduced more than one century ago?

B: Hadamard was one of the pioneers in studies of chaotic dynamics. What is now called "Hadamard billiards" is not really billiards. What "billiards" means is that you study the motion of a point particle, mechanical particle, or an acoustic wave propagating in some medium. It gets reflected from the boundary. If there is no boundary, it is not a billiard. Systems without boundaries are the simplest — they are just geodesic flows and were studied before billiards. Hadamard's fundamental work is not really about billiards. I was surprised to hear this name "Hadamard billiards". It was given by a physicist working in chaos theory. It's kind of confusing, but it's in the literature now. Unfortunately there is much confusion in giving names in chaos theory. Many people are working with billiards in applications because it is a very natural physical model in mechanics, in statistical physics as well as for light and sound propagation, in mesoscopic and in atomic physics.

I: What about some of the advances in percolation theory?

B: In percolation theory, I was only working tangentially and would be embarrassed to talk about it as I'm not an expert.

I: Are there any surprising or counter-intuitive discoveries in your research work?

B: There were quite a few. The first was right after my PhD — there I proved a theorem for a rather general class of systems of billiards. But after it was published, I realized it had some consequence which was very counter-intuitive. I published a very short paper, which is a special case of the research conducted in my PhD, and this paper had a hundred times more citations than the general paper. This result is very easy to explain.

Consider a narrow parallel beam of rays emitted by some flash light. Let this beam propagate in 2-dimensional planar region (a billiard table) with mirror walls. It gets reflected from the mirrors. Question is whether the entire region will be illuminated — that is, for all points inside the region, some ray will pass through them. If all mirrors are convex inwards — this was introduced by Sinai — the beam becomes broader and illuminates much more. But if it is a concave mirror, like a circle, it illuminates less. Therefore there was a universal understanding that if you have dispersion at the boundary, then it is strongly chaotic — it illuminates everything and you lose precision fast. In a circle, the beam of rays just goes around and there is no illumination of the central part of the circle. I considered a perturbation of the dispersing boundary by small focusing pieces; then it will still be chaotic. It doesn't sound surprising. It occurred as though there is another mechanism of chaos generated by the focusing boundary. For instance, if you take a circle, cut out a small piece by a chord and consider a billiard inside such table, then it is strongly chaotic because of defocusing: between any two consecutive reflections a beam of rays will pass through a focusing point and become divergent, like in billiards with dispersing boundary. Eventually it will illuminate the entire region. This was a real discovery — nobody thought about that, I didn't expect it either. The funniest thing is that everybody refers to this short paper rather than to the one from which it follows. This mechanism of defocusing was found in many other systems. It revealed that chaotic behavior is much broader phenomena than people thought before.

I: Are there some strange physical consequences of that kind of behavior?

B: There are; actually, experimental physicists in many physical labs constructed this type of devices and studied this phenomenon. This is what I like. When you come up with some clear examples, physicists go to their labs and build real physical analogs of these "purely mathematical" toy models.

Another example is from a totally different area, in psychiatry. My first papers were published in genetics journals. At that time, there was a theory by an outstanding geneticist in the Soviet Union that any hereditary disease is confined to genetic families (consisting of all ancestors); that is, roughly speaking, if two persons are carriers of a hereditary disease, then there is a high probability that they are relatives (in the genealogical tree). There was such a strong claim based on some computations in genetic populations. It is a fundamental problem for the organization of health care. Dealing with it, I introduced a new class of models in population genetics, which was called hierarchial models of human population. Models that were considered before assumed that the population is mixed, people get married randomly or there are several such populations with (horizontal) migrations between them. But we see that people from small villages usually migrate to cities, from small cities to bigger cities, and so on. Migrations in the opposite directions are essentially negligible. Of course, there are not so many layers, roughly speaking four or five even in the developed countries. However, this hierarchial structure is very important and changes the behavior of the population very essentially. My computations for such hierarchial populations gave the distribution of the special genes that are carriers of hereditary diseases.

Several years later I was at a conference in mathematical physics, and a physicist from Germany asked me whether I had a brother. "Yes, I have a brother," I said. "Oh, your brother is working in population genetics," he said. "No, it's me, not my brother." He was very surprised, and said that he had a friend who worked in demography, conducted experimental studies in Germany and could not explain the results and measurements, especially in urban areas. Then somebody told him about my model and everything was explained. It was exactly the same situation — there are many relatively big cities close to each other, coal miners were living in Essen, Dortmund, Duisburg, etc. Thus several big and well-developed cities are extremely close to each other, and form the rich high level in the hierarchy of migrations. The demographers there said the population structure did not fit any model, but the hierarchial one worked quite well.

I: You were already working on problems in biology and medical science long before the Human Genome Project. Have you applied your ideas to bioinformatics?

B: Actually I'm working in bioinformatics in Georgia Tech. We have a big effort in bioinformatics there. In Georgia Tech we had the first Master of Science program in bioinformatics in USA. Now we have also a PhD program in bioinformatics. You know, bioinformatics is another buzz word in a sense. I like it; it's better than chaos. But still, some people ask, what is bioinformatics? To me, it is analysis of medical and biological information in a general sense. But often people refer to it merely as the computer analysis of long molecules like in the Human Genome Project — DNA, proteins. A few years after the Human Genome Project, we know the letters but not the language; you don't know what is written by these letters. I think it's extremely tempting to bring in mathematics at this level and this is what people are trying to do. I believe that one of the major problems with biology is that there are no biologists who, like physicists, know and understand mathematics. All areas of mathematics are based on calculus or analysis. Historically, all the examples there were taken from mechanics and physics. Biology majors are not interested in calculus courses because there are no examples from their science. This is one of the major obstacles we need to overcome and this is what we are doing in Georgia Tech. We have developed new courses and now have several sections of calculus: traditional for engineering students and a new one for life sciences students. It's not a big deal. We just collect examples from biology, chemistry, biochemistry, genetics as a basis of this course. I hope that in five years' or 10 years' time, a new generation of biologists educated in mathematics will appear. A new thinking is needed.

I: Biology is changing very fast nowadays.

B: Yes, but still very slowly. Computers are now used and many people believe they can compute everything. But you should understand what you have computed. Here mathematical modeling is necessary.

I: How much of the computer do you use?

B: I don't use it myself but I really appreciate this possibility to conduct mathematical experiments. It's great. You have some idea and you can see whether it works or not by simulations. My students and collaborators use them.

I: Do you have a lot of graduate students?

B: At Georgia Tech, I usually have three graduate students. On the average, in our department, there is one graduate student per faculty member. I don't know whether three is a lot. In some other places, people have more.

I: Can you tell us something about the Southeast Applied Analysis Center?

B: Actually, it doesn't exist anymore. It was created by the Georgia Tech Department of Mathematics which became one of the leading research departments. We won a tough competition for a NSF grant with other departments in US. We were running projects and lectures for a lot of universities and colleges in the Southeast informing them about new developments in mathematics. We also had postdocs and some of them became visible researchers and won prestigious prizes. We are now trying to launch another center which will be more oriented to biology and ecology. The Southeast Applied Analysis Center was more oriented to probability and discrete mathematics. There are no more funds for this program now. In US, if there are no funds, it is just a name. So SAAC naturally disappeared.

I: The new center you mentioned is a kind of successor?

B: Yes, it is a successor. It is a kind of natural and major development for Georgia Tech where biological studies became a high priority area.

I: What will the new center be called?

B: I suggested "ABC Mathematical Center". A stands for "applied", B for "biological" and C for "contemporary mathematics" — contemporary in the sense that ABC will be more oriented to the studies of new contemporary topics like biological networks, systems biology, evolution biology, cell biology, bioinformatics, infectious diseases and ecology.

Tony Fan-Cheong

On Her Majesty's (the Queen of Science's) Service

Tony Fan-Cheong Chan (陳繁昌) is well-known for his interdisciplinary research at the interface between applied mathematics and current rapidly developing areas in image processing, computer vision, VLSI circuit layout and advanced architecture parallel computers. He is one of the few scholars with rare administrative and organizational skills which he has put to good use in advancing the image of mathematics and mathematicians in the eyes of the public and the policy makers in the United States. His boundless energy and enthusiasm for the promotion of mathematics is legendary.

He has served on the committees of scientific bodies like the Society of Industrial and Applied Mathematics (SIAM), American Mathematical Society, National Science Foundation (US) and the Lawrence Livermore National Laboratory (US). He is also on the editorial boards of numerous well-known international journals on applied mathematics and scientific computing. He has been invited to address many international meetings.

Chan

He has been Professor of Mathematics at the University of California at Los Angeles (UCLA) since 1986. He helped to establish at UCLA the Institute for Pure and Applied Mathematics (IPAM), of which he was the Institute's Director from July 2000 to August 2001. He was Dean of the Division of Physical Sciences, College of Letters and Sciences at UCLA from 2001 to 2006. He was Assistant Director in the Directorate for Mathematics and Physical Sciences of NSF from 2006 until his appointment as President of Hong Kong University of Science and Technology in 2009.

Tony Chan was interviewed on 12 December 2003 when he was invited to the Institute's program on imaging science and information processing to give a public lecture. The following are excerpts of the edited transcript of a spirited interview in which he talks about the interface between applied mathematics and other scientific disciplines like engineering and computer science and about his personal efforts for the cause of the mathematical profession.

Imprints: You were originally trained in engineering and aeronautics in the early '70s, and you quickly switched to computer science for your PhD. What made you switch? Was computer science already attracting many talented students at that time?

Tony Fan-Cheong Chan: The way I switched to computer science is due to serendipity. What happened was that I went through Form 7 in Hong Kong and I was good in math and physics. I was reading a magazine in high school about Feynman and Gell-Mann who had just won the Nobel Prizes in the mid 1960s. So I said, "Hey, this is where I want to go, to this place called the California Institute of Technology (Caltech)." I wanted to be a physicist and I applied only to Caltech. I did not apply to any other places.

At Caltech, I took physics classes. After sophomore I had to decide what to major in. I realized what I was good at is actually solving math problems. I was never able to say where the equations came from. I just cannot imagine that I was able to come up with those equations. What I really wanted to do is more practical things. At Caltech, you can do one or two things. After sophomore, either you take pure math, like abstract algebra, or you take applied math like Laplace transforms, separation of variables and things like that. So I took the second one. I graduated with a general engineering degree. But at Caltech theoretical engineering is applied math. I also took some graduate classes in the Applied Math Department: complex analysis, CFD (computational fluid dynamics) and numerical analysis courses.

When I was graduating, I had to decide what graduate school I wanted to go to. I was learning all these applied math. Most of the applied math problems traditionally come from fluid dynamics. You know the equations but nobody was able to solve them except in very simple cases. I do not know how to go from there to, say, designing an airplane. All I can do is flow over a flat plate rather than flow over a real wing. So I asked one of the professors: Joel Franklin. He told me there was a new field in which people used computers. I said, "Where do I go for this?" He said, "Stanford has this new Computer Science Department, and they have two very good people." One is Don Knuth and the other one is Gene Golub. It happened that Golub had just visited Caltech and I was at his talk. When the time came for me to apply to graduate school, I applied to many different areas. At Stanford, I applied to Computer Science. At Berkeley it was in Math. I also applied to some operations research departments. I was applying to places where I knew math could be applied.

Computer science in those days (1973) was very, very new. The Stanford Computer Science Department was only a few years old. You ask whether it was attracting a lot of talented students. I would say "some" but there was still a lot of skepticism about this new field called computer science. Remember that was pre-Silicon Valley, pre-dot-com, pre-Apple. Of course, Stanford had very talented students, so it was really a bit of pioneering spirit. People knew that this was a new area and there were a lot of new problems but people did not even know whether computer science was a real science in those days. Maybe it was a fad, maybe after 10 years nobody will study computer science. It was like that.

I: Much of your research spans different areas in mathematics and computer science. Does it require a special kind of intellectual temperament or mental outlook to venture into interdisciplinary research?

C: I think it does. What you need is an open mind. You've got to have some curiosity. You have to be interested in the context of your problem more than just the problem itself — where the problem arises and its broader impact. Not every mathematician has this interest. That is how I was driven into what I do. If you look at many mathematicians, they got interested in math because they discovered that they were good at math and problem solving. You give me a problem and I know how to do it. Just a very specific task. You look at the Math Olympiad, the Putnam exams. It is just problem solving and that is the antithesis of interdisciplinary work.

When you do interdisciplinary math, you are working with someone from outside math and you are often asked the following questions: Why can't the other person do what you do? Why do they need you? Why do the engineers do this? Why do they need mathematicians? But many mathematicians say, "I don't do the science or engineering stuff even though I could. But they can't do what I do." One of the most powerful things about mathematics is that it can extract ideas from one area and apply them to many different areas. The engineers and scientists are only interested in their own problems. They are not interested in other problems. So a mathematician can be a sort of broker. I personally have done it many times. For example, I am looking at imaging, but many of the problems, ideas and techniques came from computational fluid dynamics. When you come down to the mathematics it is really the same idea. That is one big advantage for mathematicians and I think it is very powerful. It is not just that you know the technical aspects of math better than the engineers.

I: In principle, the engineers could learn the mathematics themselves.

C: But in most cases they are only interested in their own problems. That is the difference between mathematicians and engineers. They don't get awards by looking at the broad mathematical theory. They get awards by solving their engineering problems. I have always said that mathematicians don't have a monopoly in doing mathematics. It's just that we are called mathematicians. The engineers do it, the scientists do it, the statisticians do it.

I: But mathematicians do it better, probably.

C: Well, it depends on what you need. Mathematicians, of course, do the internal structures and they look at extensions. They also do proofs. Nobody else does that. But if you look at mathematics in terms of being relevant and of having impact, I think some of the non-mathematicians are also very good at that. You can see this many times even in this workshop. Many ideas came from physicists, engineers and others. It is not a static world. Historically, many ideas in mathematics came from other fields.

I: But the original ideas that came were sort of non-rigorous. Mathematicians couldn't stand anything that is non-rigorous.

C: Right. I think it is good and desirable to be able to prove things and to be rigorous. But even that is not the exclusive definition of mathematics. I know that's how many people define mathematicians: we do proofs and other people don't. I don't agree with that definition.

I: But don't you think that mathematicians have some kind of compulsion to do things rigorously? It is in their nature.

C: But it should not be exclusive. In applied mathematics, it is often different. In pure math, of course, you cannot publish "kind of a" theorem. You know there is no such thing. In applied math you are willing to tolerate a bit more. You know something works, has a sound basis and has been demonstrated a lot. You also try to prove what you can. You trust your intuition. It is a different culture and a different mentality.

I: You have been actively and deeply involved with efforts to advance the lot of mathematics and mathematicians. This must have required much personal sacrifice of time for your research. Was there any special calling that you were responding to?

C: I wouldn't say "calling". I have not realized how much of a sacrifice it has been. First of all, time. But the other is a change in mentality. It is often political because when you have to deal with other people there is controversy and the issues are not clean. It is not just true or false, as in mathematics. You have to deal with human mistakes and broader political issues. There is no clean answer. For mathematicians it is frustrating because we want well posed problems with unique solutions. In human and political problems there is no such thing. You have to compromise, to give and take. In a way, it is for the same reasons that I do interdisciplinary mathematics. You got to look at it from a broader perspective. What we do is just part of a whole complex of human activities. How do we relate to society, to human history?

I: Somebody has got to do it.

C: Yes, somebody has to do it. Not every mathematician is cut out to do it. You have to be willing to think in this fuzzy way, you have to interact with people. Many mathematicians would think that the time used to deal with these other issues is time wasted because you can prove another theorem. It's certainly true at a personal level. I have been a professional mathematician for 30 years and I can continue doing it for 20 more years and then I will retire. But I can also do something else. It depends on whether you think you are only good at one thing or also good at another thing. I don't plan it. The opportunity arises and then I am willing to try new things. That is what I am like. I think my own interest, and perhaps my talent, is not limited to one area. I always say that if I don't try it now when the opportunity comes, I will regret it later. I am willing to take failure if it doesn't work out.

I: Not everybody is approached to do this.

C: Yes, it is part of a feedback system. So even though I don't seek it out, other people seek me out. I don't take myself too seriously, maybe I'm good with people and I don't offend people.

I: Do you see any improvement in the status of mathematics, or at least of applied mathematics, in the public perception?

C: Yes and no. Yes in the sense that you see it in the mass media. There is actually a lot of coverage of mathematics in the movies: John Nash, *Beautiful Mind, Good Will Hunting* and so on. There is a lot of awareness but if you look deeper into it, I don't think it is because people realize how important mathematicians are. Mathematicians are still viewed as a different breed. John Nash is a good example: obviously he is a genius but he is so smart that he has gone crazy. In a way, they are saying that mathematicians are just different and are not relevant to what they do. I think the big danger is that even if people are revering mathematicians, they don't know why mathematicians are doing it.

I talk to a lot of other scientists. They sometimes view us mathematicians as very, very smart people who prove theorems but are not aware of what other scientists do. They think that mathematicians are just not relevant and not part of their enterprise, not part of science. They do not know the history of mathematics. That is the big danger especially in the US where mathematics is viewed as science, not as art. In a way you can also look at math as art, I know in NUS you can get an arts degree in math.

I: Would you say that public perception of mathematicians has changed over the years?

C: No. There is more media coverage, people are more aware of mathematicians. But in terms of what mathematicians actually do, the relevance to their everyday life, I don't think it has improved. I am giving a public talk on Monday. That is my reason for giving it.

I: Could it be that mathematics is something like the software? What people see is really the hardware.

C: I have said in a front page article in *Los Angeles Times* (1977) that math never gets into the story while everybody else gets the credit. For example, in medical imaging, you have computer aided tomography. Think about it. Why "computer aided"? It should be called "mathematics aided" because when you look at the basic point — it is mathematics. But the public doesn't understand. The public equates the computer as the one that solves everything. They don't think about algorithms because they are too abstract. They think about software. Software you can see, something you can buy. Computer and software replace mathematical concepts and algorithms. Even the newspaper editors don't use those words. They only say "computer", "software".

You know, in weather forecasting, viewers say the computer using this software is doing it. That's all they say even though a mathematical concept is there. I think that this is a big danger. Mathematicians are not out there reaching the public. In order to simplify and in order to reach the public, the mass media just bypasses the mathematicians at the interface. The computer is a tool. You would never say a writing pad is a great novelist even though the writing pad is an important tool. It is the intellectual ideas that should matter, not just the tools. And what goes into the computer is a part of what mathematicians do. But that is never talked about and people don't know. That really is the problem.

I: You chaired the Local Organizing Committee for the AMS conference on *Mathematical Challenges of the 21st Century* in 2000 at UCLA. From your point of view, what is the greatest mathematical challenge of this century?

C: The idea for this conference that we called the Millennium Conference came from the then AMS President Felix Browder. It was to be like the one in Paris (in 1900). One thing you realize is that, unlike what Hilbert did, one person cannot do it anymore. There were 36 experts and in fact, they were not all mathematicians. There were some computer scientists and some physicists. I'm not a pure mathematician but I went to every single talk. It was a chance of a lifetime. One thing I realized is that the connection between the different fields is one of the strong themes that came up. The connections between analysis, number theory and geometry go back to Fermat and Andrew Wiles. The Langlands program is one of the big challenges. We haven't quite come full circle but, to me, the connection between mathematics and other disciplines is the big thing — mathematics and computer science, mathematics and the biomedical world. And there are other intellectual fields. In a way, everybody knows that this century is going be the century of the biomedical world because of the genetic revolution. So the biggest challenge is what the role of mathematics is in this. I really do not think that mathematicians have grasped this opportunity yet. A lot of other people have. Certainly statisticians have. Physicists and chemists have also gone in there. But mathematics as a field has not really come to grips with this.

I: What about applied mathematicians?

C: Yes, some of them. Even then it is not so clear because it requires learning another field. It requires a new way of thinking about new problems. In my view, you have to learn what the relevant problems are and then you ask what are the relevant techniques that you have or what new techniques you have to develop in order to apply them. That requires a change in direction. I think one of the liabilities that a mathematician has is that it takes so much time and effort in learning the tools in a certain area, especially in a very difficult field in pure mathematics in which you have invested. It makes it very difficult to change fields. But in applied math it is a little bit easier to change.

Applied math is where you extract the ideas, like PDEs, how to compute and so on. And this can be applied to many different fields. So by this very nature we can adapt. And the problems that motivated some of the techniques change through time. The typical applied mathematician probably changes, not the field, but the problems that they solve. At least two or three times in their career. In pure mathematics you don't change as much.

I: But how do you get the topologist to be interested in a problem about protein folding and this sort of thing?

C: You probably need a few leaders. You need some people who will the take the risk. There is a famous example in computer science: Dick Karp, who won a Turing Award. A decade ago, he thought that biology was going to be important (the genetics stuff) and he probably said, "I have proven myself and won the Turing Award. I am now willing to learn about biology and I want to find out." So it takes people like him to lead and then people follow the role model.

I: What is your greatest achievement in your efforts to bring public awareness and recognition to mathematics?

C: I have a ready answer for that. I mentioned this *LA Times* article. After the article was released, I got feedback and I realized the power of the mass media. When you really want to reach the public, the mass media is so much more powerful. I think more people read about my work through the *LA Times* article then the rest of my papers combined ten times. The *LA Times* is read by several million people a day. A typical math journal: if you get over 100 citations it is very good.

I: In some sense the exact nature of mathematics works against it.

C: Yes. But we are not trying to publish a theorem, we are just trying to publicize the idea. So you have to put the precision aside. Some journalists are very good at knowing what the public wants and at translating what you say. They know that if I use this word, it is too abstract and too technical. That is why I have more respect for these science journalists after the interview. I would argue with them and say, "Look, this is the right word." They said, "No, no. Say it this way and the public will understand."

I: It seems that technological advances in computers have pushed us into the direction of using more and more sophisticated computational techniques in solving concrete and real-life problems. Is this the way to go for advancing our knowledge of the universe? Could we have missed some ideas which could revolutionize science and which are basic and "idealistic" but non-computational?

C: I know exactly what you mean and I agree. But in the end the computer is just a tool. It is a very important tool and it is becoming more and more powerful, so people are using it more. But I don't think we should abandon the thought process, the ideas, the understanding. The computer is important but it is not going to solve all the world's problems. You have got to have understanding.

I: Are we over-relying on the use of computers to solve problems which cannot be solved exactly?

C: I don't think it is over-relying. It is a relatively new tool and people are exploring it. There are still physicists who think about string theory, the grand theory of everything and they don't rely on computers. I don't think we are running into any danger.

I: Would the use of computers one day shed some light on how the brain works?

C: Yes, that I believe. People are doing that. You can simulate models. What happens if the human mind were to work this way and what can it do? You can then use the computer to simulate. People in computer vision do that. But you cannot turn it into computer software. I don't believe in that.

I: You have covered so much ground and issues. Thank you for your time.

Sun-Yung Alice

Analyst in Conformal Land

Sun-Yung Alice Chang (張聖容) is well-known for her many important contributions to real harmonic analysis, geometric analysis, nonlinear partial differential equations and applications of analysis to problems in differential geometry. In 1995, the American Mathematical Society awarded her the Ruth Lyttle Satter Prize in Mathematics (awarded every two years to a woman for outstanding research in mathematics) for her deep contributions to the study of partial differential equations on Riemannian manifolds.

Born in Xian, China, she grew up in Taiwan and had her undergraduate education at the Taiwan National University in Taipei and her PhD at the University of California at Berkeley. She has taught at the State University of New York at Buffalo, University of Maryland and University of California at Los Angeles before moving, in 1998, to Princeton University where she is a full professor.

She has given invited addresses at major mathematics meetings and conferences, including a 45-minute talk at the International Congress of Mathematicians (ICM) at Berkeley in 1986, a one-

Chang

hour plenary talk at the ICM at Beijing in 2002 and an AMS Colloquium talk in 2004. In 2001 she gave the Emmy Noether Lecture of the US-based Association for Women in Mathematics. She has served as editor of several leading mathematical journals and was Vice-President of the American Mathematical Society from 1989 to 1991. In 1988, she received the Outstanding Woman of Science Award from University of California at Los Angeles. She is a member of the US National Academy of Sciences. Her life and work is a fine example of what women are capable of achieving in mathematics and has set an inspiring role model for women pursuing careers in the scientific field. Her husband Paul Yang is also her long-term collaborator in mathematical research, and they have a son and a daughter.

Alice Chang was interviewed at the Institute on 12 June 2004 during her visit to give invited lectures at the program on *Geometric Partial Differential Equations*. The following is a vetted account of the interview in which she talked about her school years, her fascination with and devotion to mathematical research, and her views about the need to encourage women of talent in mathematics.

Imprints: Was mathematics your first career choice when you were at university?

Sun-Yung Alice Chang: Yes, it was. In Taiwan there is an entrance examination for college, but I was one of the small percentages of students who did not need to take the entrance exam. I was one of the "bao-song" (包送 — literally, "guarantee send" in Chinese), the people in each high school who can choose which college to attend without taking the entrance exam. The positions are allocated according to class standing. I was ranked first in my high school. I had my college education in Taiwan National University in Taipei. In high school, I liked both Chinese literature and mathematics. In college, I decided to major in mathematics.

I: You started as an analyst but you are now interested in problems about geometry. Do you consider yourself to be an analyst first and then a geometer, or the other way around?

C: I consider myself to be an analyst first and a geometer second because of my background and the way I think about mathematics. So basically I am an analyst and now I am working on problems which are very geometric in nature. Fortunately, I have other co-authors who are more of a geometer than an analyst. So we cooperate with each other. I always think of myself as an analyst.

I: There have been many recent developments at the interface between geometry and partial differential equations. When and how did the interaction start?

C: It has a long history. The interaction between geometry and partial differential equations (or between geometry and analysis in general) is most natural. I would say it started even in the 19th century. Geometers like Poincaré already used the analytic approach to study problems in geometry in the late 1890s. There are also the geometers of the previous generation, like S. S. Chern, Atiyah and Singer, who laid the foundations for approaching problems in geometry using analytic methods. There has also been pioneering work by contemporary people like Nirenberg, Uhlenbeck, Schoen and Yau.

I: Partial differential equations are analytic in nature.

C: Yes. You first study the problem in the plane domain and then you study the problem in Euclidean 3-space and it is most natural to study it in higher dimensional space. There the curvature and the geometry come in.

I: It seems that geometry is becoming very analytical in nature. Is there any intrinsic geometry that is involved or is it just a matter of using the language of geometry? Does it involve any geometric intuition *per se*?

C: I wouldn't put it that way. I think that analysis is a tool in studying geometric problems. Geometry provides concrete examples for studying some problems at the interface between geometry and analysis. The geometric objects are concrete examples for problems in analysis. For example, you want to know what happens on the sphere — that is a concrete model. You have the analysis which is abstract analysis — convergence, weak convergence. To apply the abstract theory you need concrete examples. Geometric objects provide those examples. Of course, it is intrinsic geometry that is involved. And geometric intuition plays an important role in the approach. Analysis is sometimes a tool.

I: Geometric intuition is not something that everyone has. Is that so?

C: That is true, but on the other hand, I think that everybody has some type of geometric intuition. A problem would not be natural without the geometric intuition.

I: There are some famous mathematicians who are able to look at geometric problems and see the results even before they can prove them. This must have involved a lot of geometric intuition.

C: Yeah, yeah. If you are talking about 2-dimensional problems, maybe some people have more intuition than other people. It could also be a way of training, from their background, from the way they see things. It is true that some people have more intuition than others.

I: Is it possible to develop such geometric intuition?

C: I think so. For example, a lot of analysis problems need a lot of intuition. It's not just geometric problems that need intuition.

I: But analysis is more axiomatic.

C: Yes, it is more structured and more systematic. You are trying to derive formulas to solve a problem. However in the direction of approach to the problem — in most times, you also need intuition. You need to have some picture in your mind in both geometry and analysis. Maybe more so in geometry.

I: Have you applied your mathematical ideas to problems in physics or other scientific areas?

C: I hope to do so in the future. At the moment, no. Some of the problems I am working on are related to problems in mathematical physics. Sometimes I do read the literature in mathematical physics and see the interaction between the problems I'm working on and developments in mathematical physics. But so far, I have not applied my results to problems in that direction.

I: With the advent of computers of increasing power, and since computers operate in an essentially discrete domain, do you think that it will be necessary in the future to "discretize" geometry in order to make full use of the power of the computer?

C: I think it's the other way around. Let me explain. I'm saying that geometry has always been developed from approximation. This concept of a discrete approach to geometry has been there at the beginning. People took that approach not because of computers. For example, even in the old times people already think of the circle in terms of approximations of polygons as the number of sides gets larger and larger. The discrete approach to geometry was there before the computer. But now using computers, it's easier to take large data sets and test them.

I: Has there been much influence of the computer on the theoretical development of geometry? Is there such a thing as "computational geometry"?

C: Of course, there has been influence of computers on the theoretical development of geometry. For example, you can now construct minimal surfaces using computational methods to generate approximations and more examples. So it has a lot of influence on many areas of mathematics. But on the other hand, this computational method will never replace abstract thinking or imagination. You first need to have the idea of something that happens, and then you use the computer to test the intuition. I always think of the computer as a tool and it cannot replace the abstract thinking and intuition.

I: What are your favorite pastimes when you are not doing research?

C: I like to take walks and ride my bicycle. I like to read novels and I enjoy classical music. When I was young, I played a little bit on the piano, but not now.

I: Mathematics has traditionally been a male-dominated activity. From your experience, did you encounter extra obstacles in your mathematical career? How do you think we could encourage more female students in the university to take up mathematics?

C: It is true that mathematics has traditionally been a male-dominated profession especially at the research level. But on the other hand, I think this is due to reasons which are more — how do I describe it — social issues because women used not to have time to devote to any career. This profession could be a good profession for women. It requires a lot of thinking and you have to be very calm and patient and willing to think through things. I don't think that it should be a male-dominated profession. It's only for historical reasons. From my own experience, I think the main obstacle was that when my children were young, I felt I did not have enough time to do the work I liked to do. This is probably true for any career woman in any profession, not just for mathematics. I think this is a profession quite suitable for women in the long term. Mathematics could be done at any time. You can choose the subject you want to do research in.

I think there should be more women faculty to serve as role models. It's very hard for a woman to think that this is a possible career if the faculty in a department are all men. You do need role models. Also, I think you should encourage women. Let it be known that this profession is suitable for women. This is a problem faced by many departments in the US and around the world. It is very difficult to increase the women faculty. There are no graduate students who are women. How do you increase your women faculty? You have to reach certain standards. But this should be gradually changed with more and more women getting into graduate school and then there would be more and more women in the pipeline for assistant professors. They should be encouraged. It's a complicated issue. First, one has to understand that the intellectual abilities of men and women are the same. There must be confidence to encourage women to get into the profession.

I: What advice would you give to a beginning male or female graduate student in mathematics?

C: First, you have to be really interested in mathematics to be a graduate student in mathematics. It's a long-term commitment. You have to think that this is not temporary — you get your PhD and then... You have to be devoted and really like the subject to be a graduate student in mathematics.

The other thing is that, I feel, nowadays there is a lot of information (the world is changing very fast) through the web and conferences and so on. Maybe it's more difficult for young people to really quiet down and think through a subject more deeply — the fundamentals, the basic background. I think young people should not be pushed by fashion and should not be forced to be very quick and have a lot of publications. They should instead think more quietly and think through the foundations of the subject.

I: Of course, mathematics has a lot of competition from more lucrative subjects like economics, computer science, financial mathematics.

C: The competition is always there, of course. First, you must have a real love for the subject, so you really want to understand something. So you are willing to devote your time and take the long-term approach.

I: Also, mathematics is a very demanding subject.

C: I think if you want to do well in any subject, it is demanding. If you want to be a good musician, a good painter, a good economist or to know your subject well, it takes a lot of devotion.

I: Can you tell us something about the latest work you are working on?

C: I'm working in a field called conformal geometry. Recently, we are trying to use fully nonlinear partial differential equations to study patterns in geometry; in particular, the latest project I'm involved in is to classify a certain type of 4-manifolds up to diffeomorphism using the analytic approach.

I: Do you think collaboration is very important for mathematical research?

C: It is. You learn much faster by talking to other people. For me, I'm an analyst and, in my project, it's important for me to talk to geometers. In this case, fortunately for me, one of my main collaborators is my husband, Paul Yang, who is a geometer. I think it's very important to collaborate and talk to other people.

I: Papers written by single authors and papers written with other authors are often given different weights. What is your view on this?

C: For papers written by yourself and papers written with others, the weights should be a little different. If you have to do everything by yourself, the weight should be heavier because the speed will be slower. If you have collaborators, you could have more papers. When you work alone, you may have only one paper. If you have collaborators, you may have two or three. Even if you count it as one-half, it's still fair. However, the quality of a person's papers is more important than their quantity.

I: Knowledge is now so broad that it is quite natural for research to become more like a collective activity.

C: But I also want to say that in the evaluation system, we should leave room for people to work alone. Some of the fundamental work requires deep thinking and a long-term commitment. Some people may want to work on their own project rather than in collaboration with others and be obliged to listen to other people's opinions and be influenced by them. In the mathematical community, we should leave room for people who want to do work in their own way. Mathematical research is not just a scientific approach; the nature of mathematics is sometimes close to that of art. Some people want individual character and an individual way of working things out. They should be appreciated too. There should be room for single research and collaborative research.

Jennifer Tour

Basic Research, Hidden Returns

... Bill Gates says research "is key to our long-term position."
– Dan Richman in *Seattle Post-Intelligencer*

Jennifer Tour Chayes has made important contributions to a newly emerging and rapidly growing multidisciplinary field that straddles mathematics, physics and theoretical computer science. Her current theoretical work on auction algorithms, self-engineered networks and phase transitions in combinatorics and computer science has found applications in the Internet and the computer industry.

After her BA in biology and physics, Chayes did a PhD in mathematical physics at Princeton. After some postdoctoral work at Harvard and Cornell, she was all set for a distinguished career in academia at UCLA until one fateful day in 1996 when Nathan Myhrvold, then chief technological officer at Microsoft, approached Chayes and her husband Christian Borgs with an offer for them to join Microsoft Research. The rest, as they say, is history. In 1977, this famous husband and wife team co-founded the Theory Group of Microsoft Research, one of the most active and vibrant groups of theoretical research in industry. In addition to the impact left by the collaborative work of Chayes and Borgs with others, the Theory Group has attracted many leading mathematical

Chayes

scientists as visitors, spawning fundamental research in a way that is rarely seen in industry. This unique phenomenon has been highlighted in a recent (March 2007) issue of *Scientific American*.

Chayes is probably the most striking counter-example to the myth that women are not cut out for science or that science has no place for women. Co-author of almost 100 research papers and co-inventor of more than 20 patents, she is the Managing Director of the Microsoft Research New England lab which opened in 2008 in Cambridge, Massachusetts. Before this, she was Research Area Manager for Mathematics, Theoretical Computer Science and Cryptography at Microsoft Research Redmond. She is Affiliate Professor of Mathematics and Physics at the University of Washington, and was for many years Professor of Mathematics at UCLA. A Fellow of the American Association for the Advancement of Science and a National Associate of the National Academies, she serves on numerous institute boards, advisory committees and editorial boards. Among them are the Turing Award Selection Committee of the Association for Computing Machinery, the Board of Trustees of the Mathematical Sciences Research Institute, the Advisory Boards of the Center for Discrete Mathematics and Computer Science and the Miller Institute for Basic Research in Science, the US National Committee for Mathematics and the Committee on Assuring the Integrity of Research Data of the National Academies, the Advisory Committee on Women in Computing of the Association for Computing Machinery, the Leadership Advisory Council of the Anita Borg Institute for Women and Technology, and the Selection Committee for the Anita Borg Award for Technical Leadership. She has also served as Chair of the Mathematics Section of the American Association for the Advancement of Science, and a past Vice-President of the American Mathematical Society.

Chayes was invited by the Singapore Mathematical Society for its Distinguished Visitor Program in July 1999 and by IMS to give a public lecture at the Institute's program on *Random Graphs and Large-Scale Real-World Networks* (1 May–30 June 2006). During her visit (7–16 June 2006), she was interviewed on 12 June 2006. The following is an edited and unvetted version of this interview in which she talks with exuberance about her passion for science and mathematics, conveying forcefully the time-tested faith, if not axiomatic truth, in the inevitable and unstoppable benefits of basic research in mathematics and science.

Imprints: Your BA was in biology and physics and you waited till graduate school before deciding to specialize in one of them (mathematical physics). How difficult was it for you to make this decision?

Jennifer Tour Chayes: I have always liked many different sciences. I started out wanting to do biology, and then I did a little bit of physics — I love physics, so I decided to double-major in physics and biology. I also did a lot of chemistry as an undergraduate, one course short of a chemistry major. Mathematics was my hobby, I just enjoyed doing mathematics as science, but I didn't think of it as a profession. I thought it was fun to do mathematics. I suppose I was better in theory than in experiments, so it was probably a better idea to go into physics than into biology because at the time that I entered graduate school (in 1979) there was not a lot of theoretical biology. There was theoretical physics and there was mathematics. So I could do a lot of mathematics as well. One of the things I feel is that you don't have to make a decision to stop doing some subject in order to do another subject. I feel that I can still choose later in my career. I chose to do some computer science, and I keep thinking that maybe one day I will go back to biology. Now, more than 25 years later, there are a lot of interesting questions in theoretical biology — the field has matured so that there really is a vibrant field of theoretical biology. It has been impacted by mathematics, physics and computer science. It's always difficult for me to make a quick decision but I don't feel these decisions are permanent until you can do everything.

I: What was your area of research in your PhD thesis?

C: I proved theorems about several different systems in solid state physics. The questions were very mathematical, having to do with random statistics. A lot of what I did in graduate school, even what I still do, has to do with phase transitions — special points in a system where there is a qualitative change in what is going on in the system.

I: You applied these ideas to algorithms too. It's very surprising, isn't it?

C: Yes. Any system, when it is large enough, starts to exhibit a kind of average behavior. When I change my parameters in the system, the behavior of the system sometimes changes dramatically. That's a mathematical definition of what happens at a phase transition. The nature of algorithms changes very dramatically when you change certain parameters. A system can go from being solvable (a very efficient algorithm) to not being solvable in a short period of time. So I find phase transitions in algorithms also. At first, I was a little surprised, but I was also very excited by the connection because when I saw the connection, I had already been working on phase transitions in physical systems for 15 to 20 years. It was very exciting to me that some of the phenomena that I understood very well were manifesting themselves in very different applications.

I: Were you first attracted to problems in theoretical computer science through the mathematics or was it the other way around?

C: If it has to be one or the other, I suppose it was the mathematics first and then the theoretical computer science but, in fact, it was the physics first, and then the mathematics, and then the theoretical computer science. Finding systems that have very interesting phase transitions, I was picking them up because I love phase transitions. They seem to have some new applications for theoretical computer science.

I: You taught in the universities before joining Microsoft. Did you experience any kind of "culture shock" in this career transition?

C: Yes, I suppose you could call it "culture shock". I think it's good to experience culture shock… it was very different from the university. Things happen on a much faster time scale. One day the company is interested in one thing, and then the Internet comes along and we shift. It's a much faster time scale than that of mathematical physics. Also there are people who really care about tearing the door apart… I find all of these very, very interesting. I could choose to participate in it or I could have a more academic group at Microsoft. I feel that I got the best of both worlds. Actually when I first told my colleagues from academia that I was leaving academia to go to Microsoft, everyone of them thought I was crazy. Now many of them think that I am very lucky, but at that time almost all of them thought I was crazy because they didn't believe that I could continue to do fundamental research. But Microsoft is very interested in fundamental research. Last week, we were giving a presentation to Bill Gates on some of our research. He was very interested in the mathematical details and he asked all kinds of questions about the mathematical details. I think there's real benefit for a company to have fundamental research because you never know what is going to be important.

I: Was there any time frame for a product or objective?

C: No, we are a very theoretical group. Microsoft has a huge development organization. There are thousands and thousands of developers. They are the ones who worry about the product time scale. In research, we worry much more about trying to expand the horizons to see where the world is going to be 10 years from now, 50 years from now. I don't think it makes sense for a company to try to have its research organization compete with its development organization. We don't have a pressure to do anything on a product cycle time scale. But sometimes, once in a while, we do get things into products, and I find that exciting too.

I: If it's not considered confidential, could you tell us a little about how you came to be involved in the founding of Microsoft's Theory Group and something about its structure — for example, is it localized in one particular place? Are there many permanent members and so on?

C: The way it started at Microsoft is that I was doing what I thought was very theoretical research on phase transitions and computer science. I told the chief technology officer at Microsoft about this research. He was a classmate of mine at Princeton when I was getting my PhD and he actually did his PhD in quantum gravity, which is much more theoretical than anything I did. But he left quantum gravity and did more classical things. I was telling this to him and he was saying to me, "Oh, you should come to Microsoft. You should do this at Microsoft." And I said, "Oh, that makes no sense." Then he kept encouraging me. Finally, my husband Christian Borgs, who is also a mathematical physicist, and I looked at Microsoft and we thought that this was a company that did care about fundamental research, even though at that time there was no research lab there. We believed them when we looked at some of the other research that was being done there. The only thing was that they thought we would have a very small group and we thought we would have a larger group. So there was some talking to do to make sure that we would have a group large enough to cover mathematics and physics effectively. This was started in 1997.

I: Like Bell Labs — they have fundamental research labs too.

C: Actually, at the time that Bell Labs was getting less fundamental in research, Microsoft was becoming more fundamental. The structure of the group... we have a relatively small number of permanent members (10). We have eight postdocs who stay for a period of about two years. We have about half a dozen long-term visitors who stay anywhere from a few months to two years. Our visiting professors may come and spend a year or two years. Just like the IMS here, we have many short-term visitors (about 200 short-term visitors per year) — people who stay from one day to one month. We don't have workshops but we thought that if we were going to cover mathematics, physics and theoretical computer science and not hire hundreds of people, the best thing to do would be to bring in a lot of visitors and talk to them, do research with them and tell them to send their students to us. We have a lot of summer interns also. So it feels more like an institute than a normal research group.

I: Is there a place where the whole group is stationed?

C: We are basically stationed in Redmond. When I first went to Microsoft, we only had research in Redmond, the company's headquarters. After I was there for a few years, Microsoft opened up a few other research labs. There's a lab in Cambridge, England and in Silicon Valley. There's now a large lab in China. But the vast majority of the research is done in Redmond. Bill Gates feels it's better to have most of the researchers there so that they can interact more with the policy people and with him.

I: Can anybody apply to visit for a long term?

C: People can apply, but we don't have so many long-term visitors. If you are working with someone at Microsoft, it's more likely that you will get long-term visitor status, or if someone at Microsoft is very interested in what you are doing. Other people often come for short-term visits and if we find that there is common interest, then they come back for longer-term visits. Microsoft funds the visitors. Visitors are paid for various reasons. One of them is that some of our visitors have come up with some very valuable intellectual property for Microsoft. We find that discussing a problem — even a theoretical problem — with someone may turn out to have applications for Microsoft and if we weren't paying them, we wouldn't have a right to that idea.

I: What happens if a person develops an idea while that person is at Microsoft but doesn't fully develop it until the person has left Microsoft?

C: Well, while they are at Microsoft when they develop a valuable idea, then we can file a patent with them on the basis of what they did at Microsoft. Patents don't have to be on fully blown ideas. A patent is usually less than a paper. In an academic paper, you try to work everything out. In a patent, even if you have an idea but you haven't worked out everything, you can still get a patent for it. Now if someone starts something at Microsoft and we feel that it's very interesting, we will sometimes ask them if they would like to stay under a contract with us, maybe one day a week they develop that idea even when they go home, and we pay them for that.

I: Are you talking about patenting of ideas? That's unusual.

C: Well, you patent algorithms but algorithms are really just ideas on how to do something.

I: There's no hardware involved?

C: No, there's no hardware involved. Now, I was surprised as a mathematician, a basic scientist, when I did my first patent. You probably heard stories of patents that have very little in them, like the "one-click patent" at Amazon — people always use that as an example. But there are more substantial ideas than that even though not every detail is thought out. In fact, when an idea is very broad, it's often more valuable. Ideas in their early stages are more valuable because they are broad and then little pieces can be patented as refinements of that. For many years at Microsoft, I patented almost nothing. In my first eight years, I think I did very little patents. Now in the last year, I have done 12 patents because I happened to be working on something that has a lot of applications for Microsoft. We look at every idea and ask, "Does it make sense to patent it to the extent it is involved?" If we think that it might be used in a Microsoft product, then we just protect ourselves with a patent.

I: Does it mean that if you have patented an idea, you may be constrained not to reveal the details when you write a paper about it?

C: Not at all. That's why you should patent it. Once you have patented it, you can tell it to the whole world because then you own the rights to it. Different companies have different ways of dealing with it. There has been a lot of criticism against certain technology companies because they don't patent. They just keep secrets. That's very hard on their scientists because then their scientists are not able to publish and not able to talk to people and not able to be scientists. For us, we make the decision. We look at something and each individual makes his own decision. No boss ever tells them. If you think this is useful for the company, then you patent it. Sometimes the day before I submit a paper,

I would give something to the lawyer and say, "File a patent on this before tomorrow because I'm submitting this paper to a workshop tomorrow." This allows you to pass it to anybody because your rights to that idea are protected. In fact, it gives you much more freedom than being secretive about it.

I: Have any of your patents brought in any personal wealth?

C: It's hard to tell what the direct relationship is. Most of the patents I have done are very recent and some of them have to do with new ideas on the web. I think that some of those ideas are valuable to the company. It surprises me. I didn't think I was going to do math that is going to be passed onto the bottom line. It shows that it makes sense for a company to have a basic research outfit because you don't know what's going to be important. It turns out that algorithms are important and mathematicians are good at doing algorithms.

I: So it benefits the company more than you personally? The company has the rights to the patent.

C: I feel that it's fair. I don't teach. I get to travel. I'm well-compensated. I have freedom to invite collaborators. For me it's a very good trade off. I love my life. I'm happy that I'm able to do something that is worthwhile to the company to justify the expenses.

I: You mentioned that there are only 10 members. Are they mostly mathematicians or physicists?

C: Mostly mathematicians and theoretical computer scientists.

I: What about logicians?

C: We don't have any logicians, but we do have some combinatorialists. Certain parts of combinatorics are very close to logic. In Microsoft Research, there is one group that was started by someone who does logic. He's now doing other things but he was a logician. He's James Gorbit. He came from the University of Michigan and he started a group on abstract state machines but he did logic for many years. So there were people who did logic at Microsoft.

I: Any plans to get a logician into your group?

C: We try to get smart people into our group. If there's a brilliant logician, then we'll hire a logician. If there's a brilliant topologist, then we'll hire a topologist. I think it's much more who the person is, the quality of their work, rather than the subject because what people work on changes.

I: Does the Theory Group select only problems that are immediately relevant to computer software and technology?

C: No, absolutely not. We do basic research just like what you would do in a Math Department or a Computer Science Department or a Physics Department. Sometimes we would talk to people in products and if the problems that they have are interesting mathematically, then we will look at those problems. We are really motivated by basic research.

I: Is collaboration more important or is the individual encouraged to work freely according to his or her own interests?

C: Definitely, we want everybody to follow their own interests. On the other hand, we really do like collaboration. So you might have a couple of people going off doing something that doesn't have anything to do with what everybody else is doing. But when we hire people, we try to hire people who like to collaborate because we feel that there is a lot to be gained by collaboration. Also, we have many visitors who take advantage of our visitor programs and we expect people to collaborate a lot. I think that in trying to cover so many different fields with a small number of people, it's important to have people who like to collaborate because they can then bridge the gap.

I: Which do you think is more decisive for advances in computing — a conceptual revolution in theoretical computer science or a technological revolution in computer hardware?

C: I think they actually go hand-in-hand. As in many other sciences, when you work in experimental sciences, you see that there is an advance in experiments and then there is an advance in theory and then there is advance in experiments, and they go hand in hand. You find the theoreticians being inspired by the changes in the hardware and the people who build hardware inspired by the software revolution. Something has come up in hardware now — Intel and some other companies have produced the so-called "multi-core" chips. In such a chip there is a potential for parallel computation. That requires a true revolution in software. Intel was really like Microsoft in coming up with new software for multi-core chips so that people will want to buy those chips. Here is a hardware revolution and now we have several groups at Microsoft trying to figure out how to use these multi-core chips. There are a lot of very interesting theoretical problems and they ask people from the Theory Group to come to talk to them. Also there are other changes that are brought on by changes in software. A lot of the revolutions we have seen in computer science recently have been done by theory people who work at search engines. The two young guys who started Google were theory students at Stanford and they came up with the first algorithm for search engines. If you look at the whole field of web hosting, which is how to deliver content rapidly and is very important to the web, the web would not be as big as it is were it not for web hosting. Size would be going down left and right whenever people try to log on to them. Akamai, the biggest web hosting company in the world, was formed by a theoretician Tom Leyten and his students. These were revolutions in theory and software, and made hardware follow along. We are building this whole structure of the Internet on the web because there were some software ideas.

I: You mentioned the chip by Intel. That is a technical achievement. Was it necessary to do that? That's just making it smaller, isn't it?

C: Well, it's not just making it smaller. Having many processors on one chip, the multi-core chip is qualitatively different from the old chips. Within each multi-core chip, you can do parallel computation. It requires a completely different kind of software on a machine language level. It is a real revolution, and I think that as we learn how to take advantage of that, we will find many incredibly new applications, just like now we learn to take advantage of increased inexpensive storage. We come up with voice applications and video applications to take advantage of Moore's Law — the increase in storage and increase in computability.

I: Do you think there is an intrinsic limit to computing power, either theoretically or technologically?

C: It would be interesting if there were a theoretical limit, something like a Heisenberg uncertainty principle for computation. It is certainly true that we will never be able to stay on more bits of information than the number of atoms in our universe. At a certain point we are going to be limited at the atomic scale. If you try to think of something along the Heisenberg uncertainty principle, you might think of the limitation in speed. However, with parallel computation, which is so much faster, and quantum computation which is a kind of parallelism, I'm not sure if there is an intrinsic limit beyond the atomic scale. That's a very interesting question.

I: What about quantum computers?

C: We actually had for many years in our group some people working on quantum computing. They have now spread out to form their own larger group. I think it's a very interesting idea. The error-correction aspect of quantum computing is the most challenging aspect. Mike Friedman, who is within our group for many years and has now formed his own quantum computing group, is working on a different model for quantum computation in which you build the physical system so that it doesn't generate errors and so that it automatically corrects for errors. He's working with experimental physicists who are trying to build these things. Nanoscience in computing is also very fascinating. There are a lot of experimental advances in nanoscience and theoretical advances in quantum computation that will help us with our computing power in the future.

I: Do you know whether there is anyone who has built a prototype of a quantum computer?

C: I know that there are some quantum gates that people have made, but unfortunately those are the ones in which errors have to be corrected in the gates. So they have very limited power at the moment.

I: So the quantum computer is more like a dream rather than a reality.

C: Quantum cryptography, I think, will be used before quantum computers. It is actually a rather promising method of cryptography.

I: What about biological types of computers? Has anybody come up with anything like that?

C: Using DNA and things like that? There are a number of people working on them. I sense more excitement about them a few years ago than I sense now. I think that there are some limitations to those things. I think they can help us possibly in the next few generations of micros. Beyond that we need something more than biological computation.

I: You are also working on auction theory. Are you more concerned with the optimal algorithms for auction strategies rather than with auction theory *per se*?

C: I'm working on algorithms for auctions and for game theory in general. I'm looking at algorithmic game theory. It's a very interesting field and that's the field in which I've been filing a lot of patents. There's a lot of interest to Microsoft — very much an area in which we're competing with Google and Yahoo and some other companies. I think there are fascinating questions there. You have to come up with methods for dealing with auctions very quickly. Whenever you put a search term into a search engine, there is an auction that takes place in a millisecond. You don't even notice the time, but all of these ads that appear on the web site of the search engine are a result of an auction having taken place when you enter that term. So you need very efficient auction algorithms. Our group came up with some methods that help to prevent click fraud by coming up with algorithms in which we understand what the incentives are to commit click fraud and getting rid of those incentives.

I: What is click fraud?

C: Click fraud occurs when, for example, you are one book seller and I am another book seller, and we are both putting ads on a search engine under the term "book". Now you don't pay for your ads unless somebody clicks on your ads, and I don't pay for mine unless somebody clicks on mine. So if

I go and click on your ad, then it costs you money. But I'm not a real buyer or a potential buyer. I'm just trying to run my competitor out of business. There are a lot of problems with click fraud, but we came up with certain algorithms which get rid of a lot of click frauds. There are a lot of interesting problems like puzzles, and it's really a lot of fun. In fact, just last week we were showing some of that to Bill Gates. It's of interest to Microsoft and it's also nice mathematics.

I: Does this type of problems generate very theoretical mathematics?

C: Yes. In auction theory, if you try to auction different items to different people who want different bundles of items, those kinds of auctions are very complicated mathematically. They are called combinatorial auctions, and the number of combinations blow up very rapidly. There are fascinating deep theoretical questions — very difficult mathematical problems, NP-hard to approximate. We were also showing Bill some of the answers to those problems, which are still at the theoretical level, but it is important that we try to understand them.

I: What about applying those things that you are doing to economics?

C: Yes, we are working economics into all of this. We are doing algorithmic game theory, which brings together computer science and economics. Actually, we have had several good economists as visitors. We have been talking to them a great deal because I think there is very interesting mathematics there.

I: Have you ever thought of going back to biology?

C: I have thought about it actually. I talked to some biologists about it. There are several areas, all kinds of things in network theory, pathways to various enzymes that are close to the network questions that I'm working on in the context of the Internet and the World Wide Web. The state of diseases is certainly a biological question. There are all kinds of fascinating data-mining questions when you look at the genome. If we could use those data more efficiently, there is no question we would have cures for a lot of the diseases that plague us. I'm definitely thinking about that. I certainly want to go back to biology before my career ends.

I: One final mundane question. Did you face any kind of barriers generated implicitly by the "traditional gender mind-set" when you first joined academia or industry?

C: The first thing is that I try to ignore it. I try not to pay attention to that. I think I became more and more aware of them when I began to have students and when I began to have to make important judgment, because then I study the facts of the barriers in other people, for other people. I think there are two types of barriers: one is that there are a few people, not too many, who don't think that women are cut out for science. The president of Harvard made some very incorrect and politically stupid comments about that. I would have a pretty easy time dealing with that because if somebody thinks I'm stupid, I very quickly show them that I'm not stupid. If someone thinks you are stupid and you are not, it makes them very foolish. That is very easy to take care of. You just do very good work and no one can question that. There is another aspect of it, which is the leadership aspect. Are people comfortable with women leaders? I think that that happens in every male-dominated field. Over the years you have very confident women who take on leadership roles. After that happens, then you see changes. The changes are brought about by individuals who go in there and do such a good job that it's a moot point.

Carl

On Wings of Splines

Carl de Boor made fundamental contributions to the theory of splines and numerous applications of splines that range from highly efficient and reliable numerical algorithms to complete software packages. Some of these applications are in computer-aided design and manufacturing (of cars and airplanes, in particular), production of typesets in printing, automated cartography, computer graphics (movie animation, for example) and signal and image processing.

He has given numerous invited talks at scientific meetings throughout the world. He has served as editor of leading mathematics journals. He has received numerous honors and awards, among them the Humboldt Research Prize and John von Neumann Prize. He is a member of the National Academy of Sciences, USA, the National Academy of Engineering, the Academia Leopoldina (Deutsche Akademie der Naturforscher) and the Polish Academy of Sciences, and a Fellow of the American Academy of Arts and Sciences. He was Professor of Mathematics in the Departments of Mathematics and Computer Science at University of Wisconsin-Madison from

de Boor

1972 until his planned retirement in 2003. He remains there as Emeritus Professor and continues to be active in research as a member of the Wavelet IDR Center.

He was interviewed on 16 August 2004 when he was at the Institute to give a plenary talk at the international conference on wavelet theory and its applications. The following is a vetted and enhanced version of the transcript of excerpts of the interview which gives a frank account of the serendipitous path from Hamburg to Harvard, followed by first contact with splines (in a research laboratory in Michigan) that soon took off into an exciting world of path-breaking discoveries and immediate applications.

Imprints: Your early university education was at Hamburg and your PhD was from Michigan. Could you tell us something about the way you went from Hamburg to Michigan?

Carl de Boor: It's definitely a story of accidents. I moved to Hamburg in 1955, and I met there (as the result of a further sequence of accidents) an American girl to whom I got engaged in '58. Her father was a Professor of Political Science at Harvard, and he arranged for me to come over to Harvard for a year, in '59. My future father-in-law had known the Birkhoffs, both the older and the younger, well and had proposed that Garrett Birkhoff might give me a research assistantship and had mentioned that I had worked as an "assistant" to Collatz who was at that time a main figure in numerical mathematics, at least in Germany. Now, I had indeed worked as a teaching assistant for Collatz and even done some calculations for him. But Birkhoff misunderstood "assistant" and thought that I had been an "Assistant". In the German system, an "Assistant" is someone close to a PhD. So, Birkhoff was very happy to give me this job. But it became clear very quickly that I was not at all qualified for it. Birkhoff was a very kind person. He did not kick me out, but he gave me a good problem, and I have never worked so hard in my life, trying to produce at least something.

During that year, I decided that American university life was freer than German university life at that time. So I decided to stay, and I got married. However, I couldn't support my wife on this research assistantship. Birkhoff was then consultant to General Motors Research in Warren, Michigan. He persuaded them to give me a job. So I worked at General Motors Research and even ended up writing some papers there. But my colleagues with PhDs were much freer in the choice of problems and were much better paid, of course. So I decided I should get a PhD too, and the nearest good university was the University of Michigan at Ann Arbor.

I: Was Garrett Birkhoff at any time interested in numerical mathematics?

dB: Birkhoff was, in fact, interested in many aspects of mathematics. He was trained as an algebraist. He did various things in algebra, universal algebra, and he wrote a book on lattice theory — he practically invented lattice theory. He was also interested in applied mathematics and numerical mathematics. He wrote, for example, a book on the numerical solution of elliptic PDEs (with R. E. Lynch). Maybe his interest in numerical mathematics started in the war years — as it did for many mathematicians then.

I: He gave you some problems to work on?

dB: Yes. He was at that time working on problems in fluids. He had written a book (with Zarantonello) on *Jets, Wakes and Cavities*. He was looking at specific problems. I had to work out 2-dimensional flows over or under an obstacle, like flow under a sluice gate. I produced some numerical results which ran counter to the perceived wisdom at the time.

I: Was your PhD connected with that work?

dB: Not at all. My PhD was totally different. When I came to General Motors, they had just started to use computers in order to represent car surfaces mathematically. The notion was that once you had a mathematical description of a car surface, then you could use computers to generate cutting paths for numerically controlled milling machines to cut the dice needed for forming or stamping that surface in sheet metal. Computers had just become powerful enough for this to be feasible. When I came to General Motors, they had started using splines to represent curves and surfaces. My thesis was something that came to me as I was thinking about improving on what I found there.

I: Who was your PhD advisor?

dB: Robert Bartels. He was the only numerical analyst at the University of Michigan at that time. He was also running the Computing Center there. But, in a way, I learned perhaps more from Birkhoff because of the many interactions with him at General Motors, and from Bob Lynch, a Birkhoff student there, and from John Rice — I wrote papers with each of them there. My thesis, the topic and the writing, did not have any real input from anybody else. But I'm sure that Collatz also had some influence on me, because I had learned from him when I was in Hamburg. I did not have an advisor in the sense that I went to him every week and he would say, "That's okay" and "What's the next step?"

I: Was your PhD work crucial in shaping your future interests?

dB: Yes and no, in the following way. My thesis has to do with the use of splines in solving ordinary differential equations, and it was really looking at what is now called the projection method. You project the equation onto a finite-dimensional space and in this way get a finite-dimensional problem which you then solve. What was new in this thesis was that I pushed this point of view of projection. Second, I used not the standard functions that people used to use (people like Galerkin or Ritz used polynomials) but I used splines, especially B-splines. Unfortunately for me, I finished the thesis in 1966, and in that very year there appeared an English translation of a book by Kantorovich and Akilov which also dealt with projection methods, and in 1966 there also appeared the paper by Schoenberg and Curry concerning B-splines. So I felt scooped, and I did not publish my thesis. But it shaped my thinking because in the thesis I realized that B-splines (these are splines of minimal support) really are the right tools for understanding and working with splines. Of course, this was understood by other people before that — Schoenberg, who invented them, understood that. It became clear to me then and it colored what I did for the next 20 years.

I: What does the "B" stand for?

dB: "B" stands for "basic" or "basis". Schoenberg called them "B-splines". If you take any space of piecewise polynomials with a certain number of continuous derivatives across junction points, any such space has a basis consisting of these basic splines (splines of minimal support).

I: How much was your theoretical work motivated by problems in other disciplines in science and technology?

dB: I like to have what I do used by others. I like that very much. But what really turns me on is when in this mess, this complicated situation, I can see something simple, that it all actually comes down to something very simple. Initially a problem might have come to me because I was interested in some applications or because I like to look at problems with some applications. But once I get intrigued by it, it doesn't matter any more where it came from although I'm very pleased that people use what I do.

I: Do you seek out problems in other fields?

dB: I never have been a person to look around for problems. There are always more problems than I can do. You listen to a talk and there is a problem. I don't actively go and talk to physicists and say, "Please give me a problem." No, but I do listen more carefully to a problem that I see has some uses. If it has no use, it has to be very intriguing.

I: Did the computer play any significant role in your discovery of theorems or proofs?

dB: Well, first of all, without computers, there would be no full spline theory today. Spline theory really developed in the '60s because only then could the computer make use of it. There was then some pressure to understand better these piecewise polynomials. So, in that sense, most of what I have done has been motivated and used by computers in a central way. But these days, I also use a computer simply because it is a wonderful tool. I work on very practical things, like representing functions or solving functional equations. For anything that I wish to prove or try to understand, the computer readily provides examples. It's an integral part of my research work. I travel with a laptop and use it all the time.

I: Is there any particular discovery or result of yours that gave you the greatest satisfaction in your research career?

dB: I have had this wonderful feeling of sudden insight only a few times in my life, but I remember every one of these moments. I can taste them even now. For example, finding the dual functionals for B-splines, realizing that the recurrence relations for B-splines, which I had come across earlier, could actually be used for the stable evaluation of splines, seeing the final step in a proof that Allan Pinkus and I made up for conjectures of Bernstein and Erdös, seeing the mathematical reason for the superconvergence numerically observed by Blair Swartz, seeing the Courant hat function as the shadow of a cube, i.e., as a box spline, etc. You suddenly see, and every time I think about these moments of insight, I'm pleased all over again.

I: Do you know whether any of your discoveries have been directly used in industry?

dB: I have to smile at that question. Three weeks ago, I was at a meeting, the annual meeting of SIAM (Society of Industrial and Applied Mathematics) in Oregon, and one of the invited speakers, Thomas Grandine, gave a talk entitled *One Day in the Life of Splines at Boeing*. In this talk, he made the point that the B-spline recurrence relations that I mentioned were used at Boeing, by his estimate, five hundred million times a day.

The Fortran programs I wrote at General Motors in the early days were still in use there in the late '80s, as I discovered when I was back there as a consultant then, and may still be in use there buried in some big code today.

I: Could you briefly explain to a non-specialist the difference between approximation theory and numerical analysis?

dB: I know I mentioned these two terms in my CV. Maybe I can explain them along these lines. What I really do is work with piecewise polynomial functions or splines. You can use these functions to represent information, say to represent some function, curve or surface. To the extent that you then worry about how well you can approximate a particular function, or class of functions, by those splines, you are doing approximation theory. But how you approximate a function depends also on what you know about the function. If the function is given to you only implicitly, as the solution of a differential or integral equation, then you are solving functional equations numerically. When you develop and analyze those numerical procedures, you are doing numerical analysis.

I: I believe that approximation theory has a rather long history.

dB: Both have a long history. You might say that both started with Newton, with polynomial interpolation. Approximation theory proper maybe started in the 19th century, with Chebyshev, with his characterization of a best uniform approximation. Then there is Weierstrass who showed that any continuous function can be approximated arbitrarily well by polynomials. But it is Bernstein, in the early 20th century, who really developed the theory, characterizing the rate at which a function can be approximated by polynomials. On the other hand, numerical analysts think of Gauss as an early contributor (think of Gauss elimination, least-squares, and Gauss quadrature) and, by the beginning of the 20th century, people were finding numerical solutions to partial differential equations in systematic ways.

I: Numerical analysis started even before computers came in?

dB: Yes, definitely. Scientists have to find solutions to the models that they make of the world. They have no choice but to compute and they had to be very clever in this when they could only use pencil and paper.

I: How has applied mathematics changed since the early years of your research career?

dB: I don't know that much about applied mathematics. Some people have a global vision, they see their field in some more general context. I'm very much of an "opportunist". I see something interesting, I go for it. I don't have long-range plans. I don't worry about what's going to happen 10 years down the road. I follow what I am intrigued by. So, how has applied mathematics changed? The computer for sure has totally changed it. Before, you had to worry very much about formulating models in such a way that a good approximate solution could be hoped for. These days you are much freer to formulate a model. You can have a very complicated model and still hope to compute good approximate solutions.

I: Do you think that the computer has, in some sense, not encouraged conceptual development?

dB: Well, it is true that even some pure mathematicians these days behave more like physicists in the sense that they can explore problems experimentally, by computations. Certainly, numerical analysts now work on complicated problems without being able to prove that the methods they are using are appropriate or effective. They have to come to terms with the fact that they may not be able to prove their results in a rigorous sense.

I: Do you consider this to be a positive development?

dB: Very much so. The more freedom there is to find out something, the better off we are. Of course, we are mathematicians, so ultimately, we do try very hard to prove that what we see experimentally is actually so.

I: Mathematicians also like to create theories. If you have a lot of information being churned out by the computer, …

dB: … then mathematicians are all the happier. I think mathematicians are always trying to make order out of chaos, trying to see what is really going on and what makes it go. With the computer generating all this experimental evidence, I think mathematicians are in their element. I think having the computer is very enriching.

I: Do you have any predictions or expectations of the directions in which approximation theory and numerical mathematics will be moving in the next 10 years?

dB: I might guess that approximation theory will concentrate on the efficient representation of information, but I really have no idea, nor do I feel badly about that. As an example, in 1985, certainly nobody in my area would have predicted the onset and influence of wavelets. There were, at that point, some experts who knew about them, who knew a lot about them. There were even people in numerical methods who knew about the idea of multiresolution. Still, when wavelets hit approximation theory and numerical mathematics, it was a real surprise. So, for all I know, another such fundamental change is just around the corner.

I: What sort of advice would you give to a graduate student in applied mathematics who wishes to get started in research?

dB: First, get a good teacher. If a student does not know enough to choose a good teacher, there is no hope. Also, it doesn't matter so much what the student does or chooses to do — the teacher is there to help — but the student must feel passionate about it, must really want to do it. And then the rest, assuming that the student has talent, will not be a problem.

Persi

The Lure of Magic and Mathematics

Persi Diaconis is perhaps one of the most unusual mathematicians of our time. After studying the violin when young, he switched at the age of 14 to magic in which he had a successful and colorful career for almost 10 years, and then, at the age of 24, he made another decisive switch to mathematics.

He has made numerous contributions to mathematics, statistics and probability and is editor of many well-known journals. He has also used his expertise in mathematics and magic to investigate claims in parapsychology. He has been an invited lecturer at important meetings, notably as Wald Lecturer of the Institute of Mathematical Statistics, Gibbs Lecturer of the American Mathematical Society, plenary speaker at the International Congress of Mathematicians and Von Neumann Lecturer of the Society of Industrial and Applied Mathematics. He has won many prestigious awards and honors, and is a Fellow of the American Academy of Arts and Sciences and a member of the National Academy of Sciences, USA. He is now Mary Sunseri Professor of Mathematics at Stanford University and holds joint positions at the Mathematics Department and the Statistics Department.

Diaconis

Persi Diaconis was interviewed on 20 August 2003 when he was at the Institute as an invited speaker for the program *Stein's Method and Applications: A Program in Honor of Charles Stein* held from 28 July to 31 August 2003. In the following edited transcript of the interview, he talks about his two loves (magic and mathematics) and the excitement of research.

Imprints: You started to take up mathematics at a comparatively late stage of your life. What made you make that change when you already had a successful career as a magician?

Persi Diaconis: At 24 years old. I really don't know why I went into mathematics. It seemed to be an esoteric subject to me when I was young. It was the Sputnik era — if you get a PhD you could get a good job as a professor (it's still the case, I think). Maybe that's the reason.

I: There must be some connection between what you did before and what you did after.

D: There was some connection in the sense that I knew Martin Gardner who was a wonderful writer of popular mathematics. He put some of my early magic tricks that were mathematical into the *Scientific American* and I was very happy about that. I didn't really know any other mathematician, but in the end I felt there was some similarity between mathematics and magic.

I: You already used mathematics in your magic tricks?

D: Oh yes, there are some magic tricks that use pretty elaborate mathematics. For example, magicians can perfectly shuffle a deck of cards. I learned that if you do that eight times, the deck comes back to where it started. One of my early discoveries involved the two types of perfect shuffles, in and out. The out shuffle leaves the original top card on top, the in shuffle brings it second. If you want the top card to be in some position j say, then you express $j-1$ in binary form and use the bits of $j-1$ as instructions for the shuffles. That's how you get the top card to where you want it to be. Well, that's a mathematical discovery. There are all kinds of tricks that use mathematics, most of them awful, but there are some good ones.

I: So you actually discovered some mathematics while you were in magic.

D: Something about binary numbers, also Fermat's Little Theorem $2^{p-1} = 1 \pmod{p}$, and some elementary number theory.

I: You mentioned before that doing mathematics is like doing magic. How is that?

D: One similarity is this: you have to solve a problem and you have certain tools that you are able to use and others that you are not allowed to use. And as in problem-solving there is the notion of elegance. The difference is that mathematicians have hundreds of years of tools whereas in magic you use whatever you can get. The similarity is especially so in applied mathematics in which the problem comes from somebody else. The chemist or biologist might have a question for you, and you don't have any ready tools. You have to start thinking about it and start using whatever tools you have or invent new ones. That's pretty similar to solving magic problems.

I: In some sense, magic is as logical as mathematics.

D: In some way, but magic is not logical enough to allow everybody to see the trick. If you think about it, magic involves principles of deception. I once taught a course on the history of deception. It was an interesting course about the history of magic and the psychology of lying and things like that.

I: Did your experience with magic give you some advantage when you took up mathematics?

D: The only advantage would be first of all that I had the ideal of inventing something of my own from my magic tricks. Also I had a great mentor (Dai Vernon) who taught me the difference between mere variation and something that is really original. Another thing is speaking before the public. It makes a difference to be able to give a talk to people. To make them understand a talk is like doing a show. You have to make them follow it and enjoy it and not just sit there thinking how smart the guy is.

I: When you prove a theorem do you try to find the best way to do it?

D: Sometimes. Erdös once had a theorem about the order of a random permutation. He and Rényi had a very long paper on the distribution of the order of a random permutation. I managed to prove it in about five or six pages, and I showed it to him and he looked at me, surprised that I was right. And I remember that.

How do you know that something that you use a lot is true? I'm opposed to the recent thing about proofs from "The Book". They are wonderful, amazing proofs but they are useless. They are not the work of mathematicians because that's not the way we work. You want to tell people how you think about your problems. If you can follow the proof why that's true, that's great. But these very sleek proofs... are only beautiful like magic tricks.

I: You don't go back to re-prove things?

D: Sometimes, but not often.

I: Do you choose to work on a problem because of its potential applicability?

D: Not necessarily. I'm always happy if a problem has an application. I used to just make up the problems on my own, or work on a simpler version of a real problem if it's too hard now. I work on mathematical problems with biologists and problems from computational group theory which involve computing with large finite groups. One of the main tools in group theory is the representation theory of finite groups. I have used this extensively but also contributed by working out a non-commutative Fast Fourier Transform.

I was recently working with group theorists to try to understand why the product replacement theorem works. One of the ingredients is that you have two giant matrices, say 100 by 100, which generate a group over a finite field. You want to know something about its irreducible representations. You take a random element from the group algebra and do something like a random walk on the matrices, and it works pretty well. You can prove a theorem about it. Well, I asked myself what can you say about the general linear group over the integers modulo 2. I'm working on problems in group theory by applying probability. You have to know a lot of group theory and vice versa. I'm fascinated by problems from group theory.

I: Is this a new approach to group theory — applying probability?

D: No, not really. Many people have used this kind of approach. Erdös has used probabilistic methods to prove that an element of a group has certain properties. If you want to know whether a group is simple or whether a certain modular representation is irreducible, there are algorithms which can be used to show that they have the required properties with probability close to one. But it raises a philosophical question: if something is true with probability $1 - (1/2^{100})$, is it the same as saying it is really true?

I: How important is collaboration with people in other fields?

D: I learn many things in that way — things that I don't know about. It's very hard to read something from the papers in a different area. But we can talk to each other and in 10 minutes you know what it is about. That's one of the great joys in mathematics — to get to talk to somebody.

There is something I once learned from Erdös very early on. We were working on a number theory problem. It came down to an algebraic topology question, and he said, "Oh, I'll just call so and so, and I'll ask him." And I was shocked because I thought that we were learning from the topology books and shouldn't we work it out ourselves, but he said, "But why? If we call him, in five minutes we'll get the answers." I thought that if we want to learn some algebraic topology, we would be better off learning on our own. Anyway, there's the question of when do you ask somebody else and when do you learn it yourself?

I: How do you describe yourself with respect to your research?

D: When people ask me what I am, I say I am a statistician. I certainly do a lot of probability, but I never had a proper course in probability. I taught many courses in probability and I worked with great probabilists who taught me. When I was at Harvard, which is a very mathematical department, we didn't really have a probability course. There was a course I took on probability given by Gleason who would just teach it without knowing the literature. And he taught it in an original way. Then I try to learn group theory, and now I know more group theory than most probabilists.

I: What is the most satisfying piece of research work that you have ever done?

D: I would say the work on perfect shuffles and random permutations.

I: What are your effective working habits?

D: Well, they have changed. It used to be that I would wake up early in the morning and work till late at night. And then 10 years ago, I met Susan. Now I wake up early before the others and work for a little while and then I have to find a place to hide. If I go to the office, there are students, the email and the secretary. So I have to find some place to hide for an hour or two. My work is now more sporadic than it used to be, but I still work at night. I did a lot of work with other people on the phone. I would call people on the phone and tell them I was stuck with this problem and asked them would they tell me about this and that. It's nice to talk to people.

I: Do you think about your problems when you are doing something else?

D: Sometimes. I love to solve problems, I love to think about them. Sometimes I just have to relax. It's a hard thing to do. Some people ask me whether I work on big problems. I just work on the next problem in front of my nose. I work and think hard about it. When I came to Singapore, I started to think about new problems. Before that, I was thinking about some problems about symmetric operators commuting with groups of transformations.

I: How much of the computer do you use?

D: I used the computer a lot in my PhD thesis. I love to think about algorithms, about better ways of computing, but mostly I leave the computing to others.

I: Have there been any surprises in the way your research work has developed?

D: Sometimes. There have been wonderful surprises. To study the perfect shuffles used in magic tricks, you have to take two permutations and look at the order of the group they generate. It turns out that they are very useful for certain parallel processing algorithms. Well, some people don't like their work to be useful. But I wrote a book about representation theory of groups and probability theory and try to understand what it means for two matrix representations to be close together. I wrote a few papers about that and had some PhDs writing their theses on that too. It's wonderful for me to have a bunch of computer scientists reading my papers and asking technical questions which I considered 20 years ago. That's amazing and really surprising. Several times, I'm worked hard on a problem for my own reasons, and then somebody else is interested in it too. Some of my colleagues screen their problems. I'm just the opposite — though I do screen my problems. If I work on a problem and I have to talk to other people or to learn some new mathematics, that's a good reason for doing it.

I: Are there any problems you would like to solve?

D: I worked hard on Stein's Method when I came to Stanford. I thought I knew about it but I didn't do very much. Now I'm thinking about it again. In the airplane, I was thinking about eigenvalues. When I came here I worked hard with Zhidong Bai and we got some work done. I don't know what problems I would really like to solve. If you think about it, wouldn't it be wonderful to solve the Riemann Hypothesis? But I just don't work like that.

I: What about problems in computational biology?

D: I tried to teach myself biology for several years. It's quite embarrassing how little I managed to learn, to tell you the truth. When I first came to Harvard, there was a young assistant professor who took one of my courses and who was involved in the university's work on the human genome. He came to me with some combinatorial problems and he tried to teach me some biology. I wrote some papers on computational biology but only if somebody posed the problem as a mathematical problem. My wife Susan really learned the biology and all the other business. It's several years of work. When it came to learning biology or learning about another part of mathematics, like unbounded operators or extraspecial *p*-groups, I had to choose, and I chose extraspecial *p*-groups. It's some internal reason.

I: If a graduate student has to choose a field of research, what kind of advice would you give him?

D: I had a lot of graduate students coming to see me about research. After going through graduate school, they should know what they want to do for their PhD thesis. (I did my thesis on Tauberian theorems in number theory — but this is not what I do nowadays.) If a student has a topic, well, let's find some problem that is relatively hard to do. Another way is to take a harder field and learn about it, sometimes I say to students, "Do what you can do best." All students are different. I remember one student in my class who was spectacular with his homework problems. People were telling him, "You're good. Why don't you read this." But he wouldn't sit down and read those things. He was a problem solver. So I said, "Okay, here are some problems. Solve them and that will be your PhD thesis."

I had another student who wanted to do something real. I didn't help him very much but I didn't get in his way. I didn't make him do work that I thought he should do. I let him do something real — character recognition to read Bengali. It's partly pattern recognition and it uses algebraic topology and curves, and it really works. I'm very impressed by that. My students are free to do all sorts of things.

I: There is a lot of pure mathematics being applied to statistics, isn't it? Like Jordan algebras.

D: Absolutely. Jordan algebras in multivariate analysis, symmetric function theory, Charles Stein's work on amenable groups. Statistics used to be more theoretical but now the computer has taken over. Statistics has become very computationally oriented. They are not interested in group theory. I'm half in the Mathematics Department and half in the Statistics Department. It's not clear what we are doing is relevant to statistics.

I: What do you see are the directions in statistics?

D: Well, very big data sets. For example, I'm working on this problem on protein folding with enormous data sets and the problem is how to simplify them and make them comprehensible — data mining. How do you adjust to the richness and the power of the computer? It's an important problem. The technology is changing rapidly. So I may not be the right person to ask.

I: You have been very enlightening on many aspects of research which we don't see from journals and books.

D: People don't talk about what it is really like to be doing mathematics. If you work in a new area like me, you go and talk like a child to somebody who has been doing it for 20 years. If you can tolerate feeling stupid, something will come out of it. When you can finally think your way out for five minutes, you will say, "It's nice, How can I have missed that?" I often think that most of us who keep going have pretty thick skins. I'm very happy that you have the IMS. It's something important to have and it's a wonderful idea. You don't have anything like this in the area. It's great for the country.

I: But we are a bit isolated. We are grateful that people do come. It's a long way away for many of them.

D: Well, you can bring people in. If you put up some good meetings and you do as well as you have done, people will come to this place. You gain some reputation for good work and it has been a wonderful conference to me. I feel really invigorated and I learn something I didn't know about new problems. Some of the youngsters are very good and I've been working hard. Everyone is working hard. It's a pleasure working. We don't know ourselves. We don't know how happy we are to be allowed to do what we do.

David

Sparse Data, Beautiful Mine

David Donoho is world-renowned for many important contributions to statistics and its applications to image and signal processing, in particular to the retrieval of essential information from "sparse" data. He is reputed to be the most highly-cited mathematician for work done in the last decade (1994–2004) — a reflection of the impact of his work on engineering and the physical and medical sciences.

He has received numerous honors and awards, notably the Presidential Young Investigator Award and the Presidents' Award (of the Committee of Presidents of Statistical Societies). He is a member of the National Academy of Sciences, USA, and the American Academy of Arts and Sciences. He has been invited to give prestigious lectures of scientific bodies, such as the Wald Lecture and the Bernoulli Lecture, and at the International Congress of Mathematicians. He has served on the committees of professional scientific bodies and on the editorial boards of leading journals on probability, statistics and mathematics.

The interview was conducted at the Department of Mathematics on 26 August 2004 when he was a guest of the Department of Mathematics and the Department of Statistics and Applied

Donoho

Probability from 11 August to 5 September 2004 and an invited speaker at the Institute's program on image and signal processing. The following is an enhanced and vetted account of excerpts of the interview. It reveals little-known facets of his early scientific apprenticeship in the primeval and almost unreal world of computer programming and data analysis of the '70s. He talks passionately about the trilogy of attraction and fascination with computing, statistics and mathematics and about the many statistical challenges and opportunities arising from the exponential growth in data collected in all branches of human knowledge.

Imprints: How did you come to be interested in probability and statistics?

David Donoho: When I left for college, my father suggested I get a part-time job where I'd learn computer programming. The employment office at Princeton sent me to the Statistics Department. I went to work for a professor (Larry Mayer) doing statistical data analysis of household energy use; this taught me to use statistical computer programs. Through this I gradually became very interested in computers and also in data analysis. At the same time, I was taking mathematics courses. I saw that a career in statistics would let me do mathematics and use computers to analyze data. By the end of my first year I was hooked; I remember that I started pulling all-nighters hanging out at the computer center already by spring term.

I: Why would you need to stay up all night?

D: In those days, computing was *much* more difficult than today. It was a major effort to translate your program into a physical form (punch cards) acceptable to the computer. It was a long wait for the computer to process your work, and then often it would just spit out something like "IEHK6040 Job Control Language Operand Agreement Error". You had to be very persistent to get things done. Sometimes it would just take all night.

I: What was the attraction?

D: Today it's all the rage for young people to do "Extreme Sports"; triathlon, bungee, and so on. The point is the sheer exhilaration of taking on a daunting challenge and prevailing. Computing in those days was a kind of extreme intellectual sport. To get over all the physical and intellectual hurdles was really an achievement. By comparison, computing today is like jogging, or maybe just a brisk walk around the block.

I: What came next?

D: Computing improved very rapidly. The next year, the Statistics Department got a PDP-11 minicomputer — one of the first Unix computers outside of Bell Labs. Don McNeil and Peter Bloomfield gave me a job: to build up all the statistical software that was needed. I had to program linear regression and graphics, and I also had to provide graphical displays from exploratory data analysis. I had to learn C (a totally new language in those days) and even assembly language programming. I had to learn all the basic algorithms for statistics and numerical analysis, and implement and test things. At times I practically lived in the computer room almost 24 hours a day. I remember very clearly the room — named after Princeton statistician Sam Wilks. There was an oil painting of Wilks, the minicomputer, Wilks' personal books, the old-fashioned teletype terminals, the fancier pin writer terminals, the tape drives, the clack-clack, buzz-buzz sounds everything made. I remember all the strange things that came with staying up late at night working on the computer in this place. Sometimes the computer users held "afternoon tea" at 3 am in the Math lounge!

I: Tell us about the computing tools you were developing.

D: The computer package was called ISP (Interactive Statistical Package). In addition to regression and data manipulation, it had all the basic tools of Exploratory Data Analysis. John Tukey's book *Exploratory Data Analysis* was just being published during my Junior year in college. Later, the software was used at hundreds of universities, especially after people at UC Berkeley took it over, revamped it, renamed it as BLISS and worked on it as their daily bread. Gradually, this got displaced by the S and R languages.

I: What were your statistical interests?

D: In those days, robust statistics — being able to cope with small fractions of really bad data — was a big deal. For my Junior Paper and Senior Thesis, I immersed myself in *Annals of Statistics* papers on robust statistics. Many researchers were interested in knowing the "right" score function to use in a robust (M)-estimator. For example, my Senior Thesis adviser John Tukey had proposed the "biweight" score function which I had programmed into the ISP software. I studied the notions of minimax optimality due to Peter Huber; you play a game against Nature where you pick the score function, and Nature decides how to contaminate the data with outliers. I formulated and solved the problem of unimodal contamination. Years later Jim Berger (now at Duke) came across an equivalent problem from a different viewpoint (Bayes Decision Theory) and published an equivalent solution.

I: I believe you had some industrial experience after college. Can you tell us something about that?

D: I lived at home in Houston and worked for the research labs of Western Geophysical on problems in signal processing for oil exploration. It was after the second oil shock of the 1970s; there was tremendous interest in the oil business in finding new oil and developing new imaging and signal processing methods. I was assigned to work on what seemed (for the time) massive imaging problems. They had to fit a linear model with thousands of unknowns and tens of thousands of observations. This was huge for 1978; they used the largest mainframe computers of the day, filling up rooms the size of basketball stadiums run by hundreds of people in white coats. Computer jobs had to run for weeks to produce a single image.

The key point was that the data were actually of very bad quality, with many outliers; since I knew robust estimation inside out, I showed the geophysicists how to do robust regression. They were very eager; it took only days for a senior researcher to take it on himself. The result, immediately, was a much clearer picture of the subsurface. Western right away wanted to make this into a product and send me to conferences to speak about it. They sent me to London for an extended stay and I made presentations to the chief geophysicist at ARAMCO and developed further ideas about image and signal processing for geophysical signals. My first scientific paper came out of that work; it studied the problem of "blind deconvolution" where a signal has been blurred, but you don't know in what way it has been blurred, and you want to sharpen it up. That's a problem of real interest today. I worked on that in early 1979.

I: Was that after your PhD?

D: No, just the bachelor's degree. I wrote a paper on blind deconvolution which I finished in early 1980, and it turned out that three papers on this appeared about the same time — one in automatic control, one in astronomy and my paper. About 10 years later, such techniques began to be used heavily in digital communications (e.g., mobile phones). A big review paper in *Proc. IEEE* referred to my paper prominently, in the second paragraph.

I: How did you feel about that?

D: I felt lucky to come to the problem as a statistician, which gave me a broader view. I noticed connections between deconvolution and robust statistics, and saw that the key point was small departures from the Gaussian distribution. In robust statistics you viewed departures from a Gaussian error term as annoyances you want to protect yourself from. In signal processing, it was exactly upside down: you could view departures from Gaussian in the signal term as gifts from heaven, allowing you to recover the signal, against all odds! To explain this, I used things I had learned from the literature of robustness in the *Annals of Statistics*. My paper could actually be read by people in digital communications 10 years later with some profit; I felt that if I had been working in a more narrowly defined subject matter, my papers would have aged more quickly.

I: How did you get your PhD?

D: I went back to graduate school at Harvard, and after finishing my paper on blind deconvolution I worked on robust estimation with high-dimensional data. I became obsessed with the idea that robustness was hard in high dimensions. Ricardo Maronna had shown that existing robust methods (like affine-equivariant M-estimates) could break down under fairly light contamination; I showed new approaches would be coordinate-free and avoid breakdown even with 49% of the data completely corrupt. Werner Stahel did related work in Zürich at the same time.

It was beginning to dawn on people in those days that we should all be thinking about high-dimensional data. In the mid 1970s, Jerry Friedman and John Tukey had made a movie of PRIM-9, a system for looking at 9-dimensional particle physics data using 3D computer graphics. I remember the thrill I got when I saw that movie as a Junior in college (the Statistics Department at Princeton showed it to students to attract enrollment in the statistics major). My advisor at Harvard, Peter Huber, had been bitten by the bug and had gotten a very fancy-for-the-day computer — the Evans and Sutherland Picture System 2 — which could be used to look at 3-dimensional objects from different angles and was shared with the Chemistry Department at Harvard. It had been developed for architecture and was being used by chemists to look at molecules. We used it to look at high-dimensional statistical data and we made movies to display our results — point clouds of statistical data, spinning around so you could sense their 3D structure. I gave talks at conferences and the high point of my talk was to show a movie of computer results; for presentation techniques, that was about 20 years ahead of the curve. I got to be a movie producer and screenwriter!

At the same time I had to write statistical software for the VAX minicomputer that hosted the picture system; I remember working late into the night surrounded by chemistry grad students and postdocs. It seems that chemists were just not as weirdly interesting as mathematicians late at night. A lot of other magical things were not so magical either. The air conditioning was always mighty cold. The operating system was not UNIX, etc.

I: How did your industrial experience affect your later work?

D: In a typical academic career you get an advisor, get introduced to a problem and a specific field, become an expert, develop linearly. My industrial experience added a whole set of other interests to my academic portfolio. I'll give you an example. One thing that I learned in industrial research is that sparsity of a signal is a very important element in doing any kind of analysis of that signal. If you look at seismic signals, they are, in some sense, sparse. The reflectivity is zero a lot of the time with non-zero values relatively rare. And this sparsity was a fundamental constraint. I saw that seismic researchers were using sparsity to do surprising things that didn't seem possible. They were solving problems with too many unknowns and too few equations — somehow using sparsity to do that. Empirically, they were successful, but linear algebra would say this is hopeless. Over the years, that paradox really stuck with me. I felt that science itself involves too many unknowns and not enough equations and that often scientists are solving those equations by adding sparsity as an extra element. This somehow rules out the need to consider all the variables at once. In the last 20 years, I returned again and again to this theme of solving under-determined systems that seem horribly posed, and yet are actually not if you think beyond linear algebra, and use sparsity. As a result, I worked frequently in applied math and information theory in addition to statistics. I have two careers, and this goes back to having worked in oil exploration.

I: Can you give some examples?

D: In seismics they have band-limited signals — that do not have frequencies outside a certain range — but they want to recover wide-band signals with all the frequencies that were not originally observed. It sounds impossible, and you can cook up counter-examples where it really is hopeless. But for signals that are sparse (most samples are zero and a few are non-zero) people were having good empirical success in seismology, and I worked with Ben Logan to prove that the problem is solvable. If you exploit sparsity, even though you only have band-limited information, you can recover a broad-band signal. Later, I considered the problem of representing a signal which was made up as a linear combination of elements coming from more than one basis, say sinusoids and wavelets. It sounds impossible, since the underlying system of equations has n equations and $2n$ unknowns, so there can be no unique answer. With Xiaoming Huo, I showed that this problem could be uniquely solved if the signal was made up of any sufficiently sparse combination from the two bases, simply by singling out the linear combination having the smallest ℓ^1-norm.

I: Although we've talked in this interview about computing and robust statistics, we haven't talked much about theoretical statistics. Yet, you have worked in this area extensively. How did you get interested in that field?

D: My theoretical immersion started early, as my undergraduate thesis solved a problem in robust statistics basically calculus of variations. I published a few theoretical papers as a graduate student, and even when they concerned "practical topics" like signal processing, they were ultimately based on things I'd learned from the *Annals of Statistics*, the main Soviet probability journals, etc. Also,

Persi Diaconis visited Harvard one year while I was there; he made it easy to believe that theory was where the fun was!

I was lucky enough to win a postdoctoral fellowship at MSRI, the mathematics institute in Berkeley. The other young visitors included Iain Johnstone, who had just joined the faculty at Stanford, and Lucien Birgé, now a professor at the University of Paris. Both were interested in decision theory — Iain, the exact finite sample "Charles Stein" kind, and Lucien, the asymptotic "Lucien LeCam" kind. I hadn't had much deep exposure to either, and my interest in such subjects really picked up. The long-run role of those interests in my career has been enormous.

My first academic job was in the Statistics Department at UC Berkeley. When I arrived there, I was equally interested in computing, data analysis and statistical theory. My career could in principle have gone in any one of several directions. I was immediately given the explicit advice "Don't get swallowed up by the computer". A certain faculty member had been spending lots of time revamping the statistical software ISP that I had developed as an undergraduate. Some faculty told me directly that I probably would go back to my computing roots and get "swallowed up" in the same way. Another faculty member gave me the advice that if I wanted tenure, I should publish 10 papers in the *Annals of Statistics*. So the message was clear: do theoretical statistics!

Peter Bickel and Lucien LeCam were very kind and patient in speaking to me about their own work and interests. These personal qualities supported me in doing what could have been very isolating work. Bickel and LeCam were also patient in listening to me explain the results of my own work, as were David Blackwell and Rudy Beran. There was a steady stream of visitors giving very interesting talks at Berkeley, with most talks emphasizing theory. It was said in those days that attending statistics seminars at Berkeley could be painful, because seminar speakers would often want to present the most challenging, abstract, and technical achievements of their life to date — meaning that some seminars would seem impenetrable. But I found them mind-expanding.

I: What's the attraction of theoretical statistics?

D: On the one hand, it's about exploring the boundary of what can be learned and what can never be learned from a given amount of measured data. On the other hand, it's about taking what scientists and engineers are inventing, and making loud claims about, and subjecting those claims to scrutiny. I sometimes feel that if we didn't have theoretical statistics, science would degenerate into a crass business of people claiming they can do the impossible from their datasets, without any fear of critical scrutiny. Finally, some ideas in theoretical statistics are just beautiful ideas, very intellectually rewarding, I think of Wald's decision theory itself, of Huber's minimax robustness theory, of Stein's insight on shrinkage in high dimensions, of LeCam's equivalence of experiments theory. You have to make a decision in life about what ideas you want to spend your time with. These ideas wear well as constant companions.

I: According to the Institute for Scientific Information most-cited website, "incites.com/top/2004/third04-math.html" you are the most highly-cited mathematician for work in the period 1994–2004, with 23 highly-cited papers and well over 1,500 citations to your work. Do you think that citation counts are important? How can statisticians increase their citation counts?

D: I'd like to emphasize that many of those papers are joint with my co-author, Iain Johnstone of Stanford. In fact he's number two in that list, close behind me. Statisticians do very well compared to mathematicians in citation counts. Among the top 10 most-cited mathematical scientists currently, all of them are statisticians. There's a clear reason: statisticians do things used by many people; in contrast, few people outside of mathematics can directly cite cutting-edge work in mathematics. Consider Wiles' proof of Fermat's Last Theorem. It's a brilliant achievement of the human mind but not directly useful outside of math. It gets a lot of popular attention, but not very many citations in the scientific literature. Statisticians explicitly design tools that are useful for scientists and engineers, everywhere, every day. So citation counts for statisticians follow from the nature of our discipline.

A very specific publishing discipline can enhance citation counts: Reproducible Research. You use the Internet to publish the data and computer programs that generate your results. I learned this discipline from the seismologist Jon Claerbout. This increases your citation counts, for a very simple reason. When researchers developing new methods look for ways to show off their new methods they'll naturally want to make comparisons with previous approaches. By publishing your data and methods, you make it easy for later researchers to compare with you, and then they cite you.

The important thing: do the reproducible research; don't worry about citations. My website has a paper on reproducible research giving the philosophy in more detail.

I: You have written that statistics is an "invisible" profession. Could you elaborate on that?

D: Many people don't even recognize that statistics exists as a discipline in academia. They are surprised when they hear that one can be a "Professor of Statistics". Statisticians, in general, don't do public relations. I think we're all too busy. There are not enough statisticians to go around. The world is flooded with data; scientists, engineers and doctors all wanting to analyze their data. Outside every statistician's office in the world, there is a line of people waiting to get in to get some help with their data. Since we are completely over-subscribed, no one is out there advertising the existence of our profession. It is a sort of secret.

I: How do you select the problems that you work on?

D: This is the problem of life, isn't it? Some problems are urgent because many people are interested in them; I like to do those once in a while because of the challenge. I often look at articles in *Science* and *Nature*. When people write articles that make a big splash, I try to understand what they did and I either criticize it or build on it. So that's one angle. Another angle is to study some fundamental area of mathematics where a breakthrough just occurred, and to trace out implications in the real world.

I: In your 2004 American Statistical Association President's Invited Address, you spoke about missed opportunities for statistics. Could you elaborate?

D: Many fun problems in computer science could be attacked by statisticians, but statisticians don't even know about these problems, partly because they are already "fully booked". Today statisticians are immersed in genomics; but there are many, many other interesting problems that are equally urgent. Go to a conference like NIPS on neural information processing. There is work on analyzing catalogs of images and sounds, problems of all sorts in signal array processing that come up in electrical engineering. There are so many interesting datasets, so many interesting problems, so many great opportunities!

I: You have worked with wavelets. How is that related to statistics?

D: Wavelet theory is a fascinating branch of applied mathematics — harmonic analysis, numerical analysis, approximation theory all come together. Studying wavelet theory you learn about representing problems, about representing signals, about representing noise.

This background is useful in statistical theory. In non-parametric estimation, everything depends on your assumptions about some unknown regression function or unknown density function. Coming merely from a background in statistics, you don't have tools to think deeply about your assumptions and how they should be represented. By learning what wavelets are all about, you suddenly understand a lot of things that were mysterious in non-parametric estimation. A simple example: often nonlinear estimators dramatically outperform linear estimators in non-parametric estimation and regression, even in problems where everything seems linear and convex and banal. Once you understand wavelets it's very easy to understand this phenomenon and extend it in many directions.

There's a wide collection of signals and stochastic processes where modeling by wavelets is appropriate — any time you have impulsive events or long memory. Many non-Gaussian stochastic models are very important in applications — remember that the Gaussian is a myth. In certain application areas such as Internet traffic, if you come with only a Gaussian stochastic process background or only a Poisson process background, you just cannot analyze the data perceptively. So knowing about wavelets widens your scope quite a bit.

The wavelet transform broadens your mind in the following way. If all you know is the Fourier transform (which every statistician has to learn in the guise of the characteristic function) then you have in your mind only a very poor collection of transforms. As soon as you have wavelet transform, you suddenly realize that there are not just two — i.e., not only Fourier and Wavelets — there are many, many transforms. The right one can depend on the data you are studying.

Finally, for many kinds of signals, the wavelet representation is sparse. That gives an impetus to the statistician to study high-dimensional parameter vectors where the vector is sparse, with relatively few big entries. Iain Johnstone and I were very inspired by this viewpoint, and it has influenced all my later work.

I: What do you think will be the forces shaping the future development in statistics?

D: Statistics is a data-driven discipline; each time someone invents a new kind of data there is always an infinite supply of new questions. Genomics is an example: microarrays came along and there were enough new questions to keep all statisticians busy. There are many new kinds of data. For example, we are now entering a world of ubiquitous sensors where there are sensors on your body, sensors in space, and everywhere sensors are talking to each other. Because of this sensor network, there will be many new questions. Another example: all sorts of data come out of analysis of blood chemistry. In proteomics, they subject blood samples to high-resolution mass spectrometry and get very finely-resolved spectra that reveal all the chemical constituents present in the blood. They hope to detect diseases early and forecast about your health. All the time we see new data sources creating enormous volumes of data with completely different structures from anything we have seen before. Basically, we need statisticians to cope with this onslaught of new data types. Each new one is going to cause a revolution in our field because you have so many new questions arising from each new data type.

I: You mentioned revolutions. Do you think there will be some conceptual revolution that will change the direction?

D: Over the last 20 years there was a shift away from an intellectual attitude, where you think very carefully before you do something, to a computational, experimental attitude where you quickly do something with the computer. At some point this will run its course and statisticians won't be able to really do much of value simply by running to the computer. Then there will be a whole bunch of new questions which arise out of dealing with these new data structures; they'll ask, "What can we learn from graph theory?" or "What can we learn from theoretical computer science?" We'll go back to a much deeper level of thought in order to make the next step. I think that's coming soon.

I: In that case, do you think there is need to relook at the way statisticians are getting their undergraduate training in order to meet the challenges you just mentioned?

D: They should be good in mathematics and computers, and really care about analyzing data. In some parts of the world, statisticians are just trained at math and they aren't interested in science. In some other parts, they learn a lot about data but are not well-trained in math. In most parts of the world, they don't get enough computer background to really push the field. It's a three-legged stool — you need all three. That's really demanding for an undergraduate education, but I just don't see any other way.

On the one hand, things are much easier these days. We used to have to work really very hard to get the computer to do anything. For even the most routine analysis, I had to write a short computer program, get it into the computer, wait for the results, and if I made one tiny mistake, I had to start all over again. It's much easier these days. On the other hand, in economically advanced countries like Singapore, in Europe and the United States, kids have so many possible entertainments that few will choose to really use their minds. It is very unlikely that more than a small number are going to look at a field and say, "Oh, this is so inspiring, I want to know *everything* about it." Kids will pursue social life and many other diversions. Plus, they'll be "cool" and sophisticated and materialistic. Finally, in a comfortable society parents may be a little afraid if their kids are too intense about study and consider it unhealthy.

Every once in a while at Stanford, I see a kid with that "look" in the eyes. I know they still exist. We get some of them in the graduate program. I'm very fortunate to have had some great students who have gone on to become very distinguished scientists in their own right. I know that more great young minds are out there. That's for sure.

Robert F.

Archway to Nobel

Robert F. Engle started his university education as a physicist at Williams College and Cornell University but switched to economics for his PhD at Cornell, specializing in the use of time series in econometric analysis. In 1982, he formulated a model, known as an ARCH (acronym for "autoregressive conditional heteroskedacity") model, to study time-varying volatility in inflation. Soon afterwards, it was realized that his model could be applied to financial econometrics. In subsequent work and in collaboration with others, he extended his model to the so-called GARCH (generalized ARCH) and GARCH-M models, and introduced fundamental concepts which have set new directions for modern econometrics. His ideas and techniques have become indispensable tools in risk management in the financial sector. For his fundamental contributions, he was awarded in 2003 the Nobel Prize in economic sciences with his collaborator Clive Granger.

Engle taught at MIT and University of California at San Diego (UCSD), and in 2000 joined the Stern School of Business at New York University, where he is now the Michael Armellino Professor in the Management of Financial Services. He is active both in academic research and in consultancy work for financial institutions. He is a member of the American Academy of Arts and Sciences and a Fellow of the Econometric Society and of the American Statistical Association. He has given prestigious lectures like the Fisher–Schultz lecture, the William Phillips lecture, the Pareto lecture, the Frank Paish lecture, the Journal of Applied Econometrics Lectures and the first Econometric Institute/Princeton University Press Lectures at Erasmus University.

He was a key speaker at the program organized jointly by IMS and the School of Economics and Social Sciences of Singapore Management University (SMU) in April and May 2004 on econometric forecasting and high-frequency data analysis. He was interviewed at SMU by Yu Kiang Leong and Kim Song Tan, the latter of SMU, on 10 May 2004. In the following edited and vetted excerpts of the interview, Engle talks about his intellectual passage from the sequestered "basement realm" of superconductivity to the gregarious, if not glamorous, world of economics and finance, how the seeds of his Nobel-Prize-winning work were planted and his views on academic research and consultancy work.

Yu Kiang Leong: Your bachelor's and master's degrees were in physics. What made you switch to economics for your PhD degree?

Robert F. Engle: I went to graduate school in physics without being sure that I wanted to continue in physics. I've always loved physics but after I started my graduate work in the basement of the physics building studying superconductivity, I decided that I didn't really want to spend my life doing research on a topic which only a handful of people would ever understand. So I went to talk to people in the Economics Department because economics is the most quantitative of the social sciences and I thought that there was a possibility of doing something useful and interesting for a large number of people. To my amazement, they were interested in having me switch. And so I did. That was in Cornell.

L: Did your doctorate work set the direction for your later ground-breaking work in econometrics?

E: There were connections. My doctoral work was in time series and some of the mathematics I learned in physics was involved with spectral representations and things like that. That was carried forward into my thesis. The work on the ARCH model was rather different although it's still time series. It was about second moment properties rather than first moment properties. It was a different class of models, but there is a relationship.

L: What led you into formulating the innovative ARCH model?

E: I was on sabbatical at the London School of Economics at that time. I was interested in a question that Milton Friedman had posed. That was a macroeconomic question. He said that he thought that the cause of business cycles was not just the level of inflation but the uncertainty of inflation. The argument is that businesses try to invest in the future. If they don't know what the price level or wage level is going to be (and there's a lot of uncertainty about it) they are likely to withhold their investments. So that will lead to a downturn in the economy. If that is really the case, then you will expect to see the uncertainty of inflation forecast changing over time and being correlated with business cycles. So that was the question I was trying to solve.

I always say that there are three inputs to the ARCH model. I brought two ideas from time series. I had done a lot of work on Kalman filtering and using predictive densities to write likelihood functions. The third input was that Clive Granger, my long-time collaborator and friend with whom I shared the Nobel Prize, had just proposed a test for a bilinear process which is a type of time series model that involves looking at the correlations of the squares of the residuals of an econometric model. One day I was on the computer and Clive came by and said, "Let's take a look at your residuals, square them, fit an autoregression." And lo and behold, that was very significant, and I said, "Wow, isn't that interesting? The data really had evidence of this sort of thing in it." But I didn't really believe that it was evidence of a bilinear model. I thought that it was evidence of something else — I didn't know what. It turned out that if I were working with this data evidence, I was able to come up with a model which could be used for convoluting volatilities to answer the Friedman hypothesis.

L: Did your physics training contribute towards some kind of insight?

E: I think my physics training was particularly important in the relationship between theory and evidence. Sometimes it starts with a theoretical hypothesis and then you look for empirical evidence. Sometimes there is empirical evidence first and the theorist looks for a model that works. I feel that whichever way it happens, that's the role the econometrician takes. He is the person who really must strive to relate the data that we see for the economy with the theoretical models to make it move. I think that econometrics is a natural way for a physicist to approach the world.

L: Would it be correct to view your ARCH model as the mother of all econometric models? In retrospect, are you surprised that it led to so many ramifications?

E: I don't think it's the mother of all econometric models. It's really the first model to be interested in volatility and it is the mother of all volatility models, but econometrics is much wider than that. So it's not at all the mother of all econometric models. I'm quite surprised how popular it was. I knew it was a good idea at that time but I thought that if econometrics is the size of a table, then the part that is interested in predicting volatility and uncertainty is only a small part of the table. But it has turned out to be very important for so many applications that are still growing.

L: Which is more important in creating models: technical mastery or intuition?

E: I think they are both important. I tend to try to prove theorems with my intuition before I get technical about them. They have to make sense to me how this could be true and then I say, "Ok, now, how can I prove it?" To me, the intuition comes first. But when I say the intuition, you have to have the technical skills to rewrite your intuition in such a way that it looks like you can understand where it fits. It's very hard to develop a new idea, because you can look at it in so many different ways. Unless you've got a wide technical background, you don't know how to begin proving the theorem. How do you phrase this theorem? You need a lot of technical background before you can even formulate the question. I'm better at intuition than I am at the technical details.

L: It seems that econometrics uses a lot of statistical theory and methods. Do you think that behind the algorithms and computations there are some fundamental economic concepts that could be subject to some kind of objective economic laws?

E: You know, when physicists talk about laws, they think about Newton or Einstein or something like that. These are inexorable laws. I don't think that there are going to be economic laws in that sense for economists because what we are looking at when we build models for the economy is the average behavior of a lot of people. By averaging you can get a lot closer to a law, but it isn't clear that it is amenable in the same way as physical laws are going to be, so I think probably not. We find general principles, tendencies and patterns that are preserved over time.

L: You mention principles, but a principle is some weak form of a law.

E: Yes, I suppose it is. When I said that, I wondered whether you would point that out. A lot of economic models are based on very strong optimizing results and general equilibrium results. Rational behavior gives you very strong hypotheses about how the world is going to be. Many of those are good descriptions of how you see behavior. So in a sense, I suppose you would think of those as economic laws but it's not that they explain things exactly. There's a lot of dynamics and adjustment that you have to make to the system that you see.

L: Modern physics deals with random behavior and so does econometrics. Do you think that there could be some physical analogies that may be useful in economics and econometrics? In particular, what are your views about quantum finance?

E: Well, I have not found that interesting — the finance theory that the physicists are doing — "econophysics", that's what I would call it. I think that it is, in an interesting way, mechanical. It tries to apply mechanical principles to economic systems and doesn't recognize that there is behavior and that it is not actually a physical model. These are agents with dual optimizing and behaving in ways that atoms and molecules don't do. So I think that while there may be interesting things that could come up out of this, I think it's not obvious that there's something very useful that physical principles can be applied to economics. I don't think that quantum mechanics has any direct implications for finance because quantum mechanics is a probabilistic statement about the future evolution of particles and atoms. It doesn't talk about the fact that in every price movement there is a buyer and a seller and somehow sellers and buyers have to agree to this kind of outcome. It isn't that one person can push the market without somebody agreeing to sell it to them. There is an optimizing character of the economy which is really not present in quantum physics.

L: You teach at the university and do research and at the same time run your own consultancy services for industry. How do you manage that?

E: I manage it by keeping them working together. So when I do my consultancy, I make sure that what I do in my consulting work is actually going to be an important part of my research, and

I have had some wonderful problems that come out of consulting projects. I think that this is a way of keeping your research focused on problems which people are interested in. I think it's important to do that but I do not like doing my consulting on things that would never end up as part of my research.

L: Do you have students?

E: I have students. I am now in the Finance Department of NYU (New York University). I have some finance PhD students and I have some economics PhD students, and I still have some students from UCSD (University of California at San Diego) whom I'm working with. I have a range of students.

L: How has the Nobel Prize affected or influenced your life?

E: In a way it changes everything and, on the other hand, it doesn't change anything. I have lots of things which are different. The press was never interested in talking to me before that. Now I have lots of interviews with the news media. They wanted to know about things that I never thought I was expert in. But I ended up talking about them anyway. I'm now more of a generalist. I've met so many interesting people from different areas of science, economics and journalism and so forth. It's fascinating about the people you meet. I meet finance practitioners. I have an interesting experience that people like hedge fund traders and so forth tell me their strategies which nobody would want to reveal in the past.

L: It's their trade secrets or something.

E: That's right, trade secrets. I don't quite know what that is but I think it's got something to do with the Nobel Prize. In that sense, a lot of things have changed. In many ways, I do my best to keep my research and my life the same as before. I'm continuing to give talks and do my research and I think I don't want that to stop.

Kim Song Tan: You were saying that after the Nobel Prize you are in a way forced to speak as a generalist in many contexts. Do you find you are more influential as a generalist in that context than as an econometrician?

E: Well, people always want to know some things like, "Is the stock market going to go down?" I don't know whether that's being influential. I haven't actually taken on any causes. Sometimes Nobel Prize winners do say, "I want to do this thing." I haven't done that yet. It could happen. I think I reach a bigger audience because I'm speaking about more general things. My general comments would be about financial management and risk assessment and that sort of things. Now I end up talking about general macroeconomic issues in the US and international issues. In about five minutes, the BBC is going to broadcast whatever I said this morning on the BBC Asia Report.

T: The reason why I ask that question is because you must have heard many times that economists these days are so focused and so specialized in their fields that they cannot deal with larger economic or even for that matter political issues. Yet economics and econometrics are part of the social sciences. How do you respond to that?

E: Well, I think specialization is natural. It's a lot to ask people to be expert in a particular area and making innovations that are valuable to the profession and to people in that particular area and still be able to speak as a generalist. However, I think a lot of times making advances in a particular area is aided if you've got a little broader interest so that you can bring things from other disciplines. You can bring stuff from mathematics, from statistics and from other areas of economics to answer problems in your particular area. So I think that some amount of generalism is a good thing anyway, but it's a lot to expect anybody in any particular area to be able to comment widely on economic issues of the day.

T: That brings us another question. Do you think that when a student wants to do economics, he should be someone who has some interest in general social, economic phenomena first, or should he just approach economics as a form of science? I'm asking you this question because you came from physics, and yet you were able to deal with economic questions and issues. I don't know whether all econometric students are of this type today.

E: I think actually it's probably true they are not. It took me a long time. When I started graduate school as a student in economics, I could do problems that someone would set but I couldn't figure out what the problems should be, what should be an interesting problem. I think it took me probably 10 years of my time teaching at MIT and so forth. I was continually trying to develop and understand this economic intuition that so many other people had taken so easily. But it was hard for me to grasp. So I don't know whether in the beginning you should expect that. But I think people should try and develop it. Of course, that's what graduate education is about. That's why you go to meetings and you listen to talks. You try to develop your economic intuition.

T: Would you go so far as to say that without it you would not be able to make it as a successful econometrician?

E: No, I don't think so. I think actually there are a lot of successful econometricians who are very narrow, technical people. They have to pick good problems. That's where you make your name. You solve a good problem and it's a kind of intuition which makes you choose the problem. I like to take problems from the world around me and figure out what actually is the nature of this problem and how you can solve it. But people who take problems from the current state of econometric research realize there is a problem here, they formulate it and they solve it. I think that's a valuable contribution.

T: A question on consultancy. Do you often find yourself in a conflicting position where the private sector tends to look for definitive answers to their questions and think there are some numerical answers to their questions and we know that it's not possible in all cases to provide this kind of answers. How do you deal with that?

E: I'm more of a tool builder. The ARCH model is a tool which allows you to study risk and a lot of consulting that I have done is not actually so much looking for answers as looking for tools. How do you build the tool that's good for measuring risk in this kind of setting? How do you build the tool that helps people form their portfolios? You build one and it helps a little bit but maybe not enough.

So then there's another one you might want to develop along the way. It's not so much getting an answer. It's advancing our real understanding and our ability to solve these problems.

T: How do you find from your experience how useful econometric solutions are to hedge fund strategy or general financial trading strategy, treasury and other types of trading?

E: I'm not so involved in trading strategy. I've avoided that because that actually doesn't ever lead to publishable research. Either it works, in which case you can't publish it, or it doesn't work and nobody cares. Even if you do publish one that does work, no one will really believe you because then people will ask, "Why did you publish it?", and it goes away as soon as you publish it anyway. I have tried not to get involved in trading strategy. But if you talk about strategies like what is the best way to forecast risk or something like that. I think those are not proprietary typically. Maybe initially you wait a little bit before you put it in the academic discipline. I try not to get involved in things that have too much conflict. Another set of consulting that I did for a long time (although I'm not doing it any more) is energy research, electricity modeling. What is the demand for electricity at different times of the day, how do you forecast that, how does it depend on appliances and things like that. This is another example of how you develop statistical methodology. People build these models for utilities and forecast what the needs will be in the future. You know it's not proprietary. There's a lot of non-proprietary stuff you can do both in the financial sector and more broadly in the industrial sector.

T: A lot of people in the finance industry these days, especially when it comes to training, tend to be engineers by training. In fact, many of them have no economic qualifications whatsoever. They seem to be doing well and form a large group in the finance industry. Do you think that in that sense finance might be closer to physics than to economics?

E: I think that finance education is typically not as quantitative as what financial practitioners require. In financial practice you need to handle an enormous amount of data, a lot of computing tasks. Finance PhDs are often not that well trained in econometrics or in computer methods and they are often trained in particular corporate finance theory or something like that. I think that finance service sectors hire a lot of engineers, physicists, chemical engineers and mathematicians because they have skills that they need and they cannot get a finance person to do. I think that academic finance is not as close to practitioner finance as you might think. In fact, practitioner finance does have a lot of economics and econometrics in it.

Hans

Efficient Markets, Random Paths

Hans Föllmer is renowned for fundamental contributions to statistical mechanics, stochastic analysis and mathematical finance. He is also known for his indefatigable energy and enthusiasm in actively promoting the applications of mathematics, especially to financial markets.

Having undertaken a broad education in philosophy, languages, physics and mathematics in four European universities, he obtained his doctorate (Dr. rer. Nat.) from University of Erlangen under the supervision of Konrad Jacobs. Except for a three-year stint in the US at MIT and Dartmouth College, his career was essentially cultivated to fruition in Europe — at University of Erlangen, University of Frankfurt, University of Bonn, ETH Zurich and Humboldt University in Berlin. At Bonn, he was professor twice, first at the Department of Economics and later, after a period of 11 years in Zurich, at the Department of Mathematics. Since 1994, he has been Professor of Mathematics at Humboldt University, Berlin.

His extensive publications cover several interdisciplinary areas. In addition to the influence of his pioneering research, he has made numerous contributions to the scientific communities

Föllmer

in Europe and elsewhere through his active involvement in scientific committees and advisory boards. For his deep and wide-ranging contributions, he received the following awards: Emmy Noether award of the University of Erlangen, Science Prize of the GMÖOR (Gesellschaft für Mathematik, Õkonomie und Operations Research), Prix Gay-Lussac/Humboldt, the Georg Cantor Medal of the German Mathematical Society and an honorary doctorate from the University Paris-Dauphine. He was also elected as member of the following scientific bodies: Academia Europaea, Deutsche Akademie der Naturforscher Leopoldina, and Berlin-Brandenburgische Akademie der Wissenschaften.

Besides giving invited lectures at major scientific meetings and universities throughout the world, he is actively engaged in the training of scientists and mathematicians both inside and outside of Europe. Among other activities, he is involved in the International Research Training Group (IRTG) Berlin-Zurich and the DFG Research Center "Mathematics for key technologies".

Since 2000, Föllmer is a regular visitor to NUS and has rendered valuable service to the Department of Mathematics and the Institute. He is a founder member of the Institute's Scientific Advisory Board (SAB) which successfully charted the direction of the Institute during its first five years. During a three-year period in 2000–2003, he helped to develop the Department's new financial mathematics program, and visited the department for short periods of four to eight weeks to advise and give courses on the subject.

It was during his visit to the Institute as a member of the SAB that he was interviewed on 4 January 2006. The following is an edited and enhanced account of this interview, in which he spoke with passion about his intellectual path from philosophy to mathematics, and gave us a rare glimpse, from the view of a pioneer at the interface of probability and finance, of the somewhat unexpected impact of stochastic analysis (an esoteric branch of mathematics) on stock markets (one of the most practical activities of an industrial society).

Imprints: Your university education seems to have been rather unusual in the sense that it was taken in many places in Germany and France. Please tell us something about it and about how you became interested in stochastic analysis.

Hans Föllmer: In the German tradition, the fact that I went to several universities is not unusual but it's quite normal and even expected traditionally. My father, for example, as a student, went to four different universities. The idea was to get to know different schools of thought in different parts of the country. In that sense, I did the normal thing. I started out in Cologne, then I went to Göttingen, and the reason that I went to France was that at that time I had already focused on one special area, and my advisor for the diploma thesis asked me to go to Paris for a year in order to learn more about it. In the meantime, he had moved from Göttingen to Erlangen, and then I joined him in Erlangen, and that was university number four — just like my father.

I: Who was your supervisor?

F: My supervisor was Konrad Jakobs. He was working in ergodic theory, and the reason that he went to Erlangen was that he wanted to establish a joint center in probability with Hans Bauer who had at the same time moved from Hamburg to Erlangen. Bauer and his students were working on the potential theory of Markov processes, and my own interest then was, in fact, closer to Bauer's than to Jakobs'. The reason I went to Paris in 1965 was that I was supposed to learn some potential theory from the sources in France — Choquet and Brelot, for example. Of course I also took other courses, and I particularly enjoyed the lectures of Laurent Schwartz and Jacques Neveu. After my year in Paris, I went to Erlangen for three years. During that time there was a lot of activity in probability. Robert Blumenthal came for a year and gave a graduate course on the book on Markov processes which he was writing with Ron Getoor. There were visitors such as Paul-André Meyer, Joe Doob, Shizuo Kakutani, Alexandr Borovkov, Kiyosi Itô and Kai Lai Chung. For us graduate students that was an exciting time.

I: Were you interested in stochastic analysis right from the beginning?

F: No, I even didn't start in mathematics at the beginning. I first started to study philosophy and literature. Then I became interested in the philosophy of language, linguistics, and I thought it would be good for me to understand how formal languages like mathematics work. So I thought I would sit in on mathematics classes, and then I got interested, and slowly I got drawn into the subject. One reason was that mathematics was much better organized as a curriculum than philosophy. Philosophy was very free floating. So I got sucked into the mathematics program and started to enjoy it. There were several occasions that I thought of going back to my original interest, but I stayed on. The reason I got interested in probability rather early, in my third year of study of mathematics, may have something to do with my original motivation in philosophy because I was intrigued by the notions of probability, entropy and uncertainty. That probably played some role in my decision.

I: Was there any single person who was quite decisive in making you work in probability?

F: The reason that I decided to specialize in probability had certainly something to do with my teacher at that time, Konrad Jakobs. He is a very impressive person and has very wide interests in mathematics and beyond. I liked him a lot as a teacher, and he immediately helped me and supported me. That probably played a role, too.

I: Was it a tradition to have broad interests?

F: Yes, that was the intellectual tradition. You were encouraged to take a broad approach and I liked that. Nowadays, it is much more focused. In retrospect, it was a luxury spending time on philosophy and so on. It would be harder to do the same thing now, also in Germany, because now there is more pressure on students to proceed quickly.

I: You taught briefly for three years in the United States immediately after your doctorate. Was it a cultural or intellectual pull that made you return to Europe to establish your career in Germany?

F: That was a very difficult decision. After one year in the States, I thought, "Okay, it was all very interesting, but, no, I really want to go back to Europe." After the second year, I was no longer so sure, and in the third year I was strongly tempted to stay. Clearly, the scientific situation in the United States was very attractive. But I was already married and we had a child, and finally we decided to go back to Germany, mainly for cultural reasons. Soon after, I had another option to go back to North America, but I also had attractive offers in Germany, and so we decided to stay. But it was more the cultural pull, not so much the intellectual in the professional sense.

I: Your research work was initially in stochastic processes, which is theoretical probability theory, and you soon began to work on stochastic problems in other fields. Why mathematical finance and not other areas like biology?

F: My research was primarily in stochastic processes for 30 years, not only initially. But I had one first contact with mathematical finance already in 1971 when I was at Dartmouth. At that time, I had an undergraduate student who wanted to write a senior thesis in probability, and he proposed to work on an optimal stopping problem related to insider information in finance. This was David Kreps, who went on to become Professor of Economics at Harvard and Stanford and to receive the Clark medal in 1989, and who is now dean of the business school at Stanford. I learned from him what an option is. That was my first contact with mathematical finance. But for a long time I continued to work in probability, on problems in martingale theory, in interacting particle systems, and in stochastic analysis. For several years, I worked on questions motivated by the interface between probability and statistical mechanics, especially probabilistic approaches to phase transitions, Gibbs measures, and large deviations. My interest in mathematical finance became more systematic only much later, in the mid 1980s. Actually, it was again triggered off by David Kreps. David spent a sabbatical in Cambridge and he came over to ETH Zurich, where I was teaching at that time, and gave a seminar related to the Black–Scholes pricing formula for options. I got intrigued and started to think about it. Then Dieter Sondermann, a colleague from Germany, visited ETH Zurich for a month. At the same time, he was doing consulting work with a major Swiss bank, and we started to work together on some mathematical aspects of option pricing. From that time on, I took a more systematic interest.

I: Did you work on a specific problem with this colleague of yours?

F: Yes, we looked at the problem of hedging financial derivatives in situations where the Black–Scholes paradigm of a perfect hedge breaks down, and we used arguments from martingale theory. By the way, Dieter Sondermann was Professor of Statistics in the Economics Department at the University of Bonn. He was holding the same position that I had held from 1974 to 1977 before I went to ETH Zurich. At that time, I had a position as Professor of Statistics at the Economics Department of the University of Bonn. That was from 1974 to 1977. In 1974, I had three options — two offers for positions in mathematics and one from the Economics Department. At that time, I decided to take the economics offer because I wanted to learn what those guys were doing. The experience of three years in the Economics Department was probably responsible for my later decision to pursue questions in mathematical finance. After three years, however, I had an offer from ETH Zurich and I thought it was a good time to go back to mathematics. One aspect of this was that in doing research with students on questions which I liked, the conditions were better in the Mathematics Department than in the Economics Department. But I never regretted the decision to go to the Economics Department for some time because it was a very enriching experience to get to know this other culture. At that time, Gerard Debreu (who later received the Nobel Prize in economics for work in microeconomic equilibrium theory) was visiting the Economics Department in Bonn for a year to work with Werner Hildenbrand.

He came with a strong group of young economists from Berkeley which included Truman Bewley (later at Yale), Mukul Majumdar (later at Cornell), Alan Kirman (later at Marseille) and Andreu Mas-Collel (for a long time at Harvard before he became minister of universities and research in Catalunya). That was a very stimulating environment, and I enjoyed that a lot.

I: How much of the field of mathematical finance has been accepted as an integral part of economics?

F: The fact that some of the Nobel Prizes have been awarded to work in quantitative and even mathematical finance shows that the field has a lot of acceptance within the community of economists. I was more concerned with the other side — how well accepted is mathematical finance as a part of mathematics? My main interest was always in questions which are motivated by the financial applications, but which also have some intrinsic mathematical interest and can be treated as research problems in their own right from the mathematical point of view.

I: Decades ago, the general public would associate financial mathematics with more commercial activities like accounting and book-keeping. Do you think that this general public perception has been significantly raised to a higher level?

F: Several decades ago, before the early '70s, I would have had the same perception. Since then there has been really a spectacular change and a dramatic increase in mathematical sophistication. Mathematical finance has become a new source of appreciation and esteem for mathematics in the eyes of the general public. In the financial industry, the number of professional mathematicians working there has become much higher than what it used to be. I think that has generated a lot of respect for mathematics within that community and also in a wider public. When the Nobel Prize was given to Mertens and Scholes for their famous option pricing formula, this was one of the rare occasions where a mathematical formula appeared on page 1 of the *New York Times*. Yes, public perception of mathematics has been significantly raised.

I: If I'm not mistaken, some kind of empirical stochastic studies of the stock market were actually carried out before probability theory was rigorously established. We know that sophisticated mathematical tools are now used to deal with problems of the stock market. Have those problems also contributed to and possibly influenced the theoretical development of probability theory? If so, could you give us some examples?

F: I think there is an interplay between direct concerns with the stock market and the development of probabilistic concepts and methods. One very basic mathematical object in probability theory is Brownian motion, which plays a fundamental role for a number of reasons. Brownian motion was proposed (not under that name) in 1900 by Bachelier in his thesis with Poincaré in Paris as a model for price fluctuations in the stock market. Thus the aim to describe price fluctuation in mathematical terms has motivated a very important step in the development of the theory of stochastic processes. From then on, the original financial input to the theory of Brownian motion was for a long time forgotten. The theory of Brownian motion was developed on its own for intrinsic mathematical reasons and it was only in the '60s that the original work of Bachelier was taken seriously again from the financial point of view. The group of Paul Samuelson at MIT started to use it systematically in the mid 1960s, and since then Brownian motion (on the logarithmic scale) serves as a benchmark model in finance.

I: Was it at a rigorous level?

F: The original work of Bachelier contained a number of important ideas. From the modern point of view, it was not as rigorous as what you would like to see nowadays. The fundamental mathematical problem of constructing Brownian motion rigorously as a measure on the space of continuous paths was only solved 23 years later by Norbert Wiener. But on the more qualitative level, some very important ideas, for example the reflection principle for Brownian motion, already appeared in Bachelier's work. It also contained a formula for option pricing. It's not the one which later became the canonical pricing formula because it was based not on a logarithmic Brownian motion but on the original Brownian motion itself, and one crucial argument for the Black–Scholes formula was missing, namely the construction of a perfect hedge.

You asked whether those problems contributed to and possibly influenced the theoretical development of probability. My answer would be "yes". I have already given the first example. The introduction of Brownian motion was motivated by the financial interpretation. Another example is the revival of martingale theory in the late '80s. Martingale theory had flourished in the '60s and '70s. The financial interpretation suddenly provided a fresh look and new questions. Several theoretical developments are due to that financial interpretation. One example is the pricing theory in incomplete financial markets. Let me explain. From the mathematical point of view, the Black–Scholes formula simply reduces to the following basic fact about nonlinear functionals of Brownian motion. A fundamental theorem of Kiyosi Itô says that such a functional can be represented as a stochastic integral of Brownian motion. In the financial interpretation, the integrand can be interpreted as a trading strategy. The nonlinear functional describes the payoff of a financial derivative, for example a call option. Thus Itô's representation theorem shows how to represent the payoff as a result of a trading strategy involving the underlying financial assets. This leads to a recipe for pricing. The initial constant which generates, using the trading strategy, the payoff may be viewed as the cost of replicating the financial derivative. This implies that the initial cost is the right price for that option. Otherwise there would be an arbitrage opportunity. That is the key to what is known as the Black–Scholes formula. From the mathematical point of view, one could say that it is simply an application of a basic representation theorem in stochastic analysis for functionals of Brownian motion.

I: Who was the first to make this observation?

F: Originally, the Black–Scholes formula was not derived by a representation theorem. It was derived by a direct argument using the Itô calculus and the solution of an appropriate partial differential equation. The full power of the representation theorem is needed if you pass from simple financial derivatives such as call options to more exotic options. Then you need the functional on the full path space. The connection to the representation theorem was clarified by David Kreps, whom I've mentioned earlier, Michael Harrison and Stan Pliska in the '80s. They recognized the relevance of previous work on the representation problem which had been done in martingale theory. It is known that the representation theorem holds if and only if there is a unique martingale measure. How to explain a martingale measure? Typically one fixes a probabilistic model for the price process, for example, a geometric Brownian motion. Such a model is specified by a probability measure on path space. If you now change the model by switching to another probability measure which is equivalent to the original one such that the given process becomes a martingale under that new measure, then that measure is called a martingale measure. To be a martingale means to behave like a fair game with respect to that measure. This notion of a martingale measure is very fundamental in mathematical finance. That the representation theorem holds is equivalent to uniqueness of the equivalent martingale measure. This had been shown, quite independently of the financial interpretation, already in the '70s and early '80s in the French school of probability; in particular by Jean Jacod and Marc Yor.

New questions which arose had to do with the fact that the martingale measure may not be unique. Then the situation becomes more complicated. The question arises: which martingale measure should one choose as the pricing mechanism. How should one construct a reasonable hedging strategy? This question leads to a projection problem for martingales and, more generally, for semi-martingales. It triggered off a new development in probability theory where the projection theory of Kunita–Watanabe for martingales was extended to semi-martingales. So that was a new version of a basic projection problem in probability which was motivated by finance.

Another example is the following. If you want to hedge the financial derivative, you may insist on staying on the safe side and make sure that there is no shortfall at the end of the day. Mathematically, this leads to the theory of super-hedging which can be seen as a new generalization of the classical Doob–Meyer decomposition for supermartingales, a fundamental theorem in martingale theory. This new version, now often called the optional decomposition theorem, was developed first, in a special context, by Nicole El Karoui and then in full generality by Dima Kramkov, a former student of Albert Shiryaev in Moscow, at that time a postdoc in Bonn, and now professor at Carnegie-Mellon. For this work he received a prize of the European Mathematical Society for junior mathematicians in 1996. This is another example where a question in finance led to a new problem in probability and triggered off a significant advance on the theoretical level.

I: The concepts and ideas are totally new?

F: The optional decomposition is definitely a new step. It is not a straightforward generalization. You can see that in discrete time. There the Doob–Meyer decomposition can be written down in three lines, but the extension to the optional decomposition, even in discrete time, takes several pages. It involves a new combination of martingale arguments and arguments from convex analysis. It is not just a technical refinement; it is a conceptual advance.

I will give you a third example. In applying arguments from mathematical finance, you usually fix a probabilistic model. Typically, there is a significant amount of model uncertainty. You cannot be sure that the chosen probability measure really describes the objective situation. One way of dealing with that is to take into account a whole class of possible probability measures. Then many new problems arise. For example, the classical problem of optimal portfolio choice translates into a new projection problem. You have to project the whole class of model measures on the class of martingale measures. In the usual case, you would simply project one single measure on a given convex class of measures.

This problem is well understood, especially if the projection problem is formulated in terms of relative entropy. The question of model uncertainty leads to a new robust version of the classical projection problem, which has been treated only recently. It has been solved last year in joint work with Anne Gundel while we were both at the IMA in Minneapolis for a program in financial engineering.

I: Does the computer play a significant role in your work on stochastic finance? Do you rely on the empirical data to shape your ideas?

F: I am staying on the theoretical side. I do not work myself on the computer or use simulations, but I follow some of the developments on the empirical side. Some of my own work is motivated by empirical work on the microstructure of financial time series. There are new modeling issues which arise. For example, if you look at financial data on a tick-by-tick basis, it provides the motivation to model the dynamics of an order book. So you do not immediately switch to the mesoscopic level of description by means of stochastic differential equations but you try to model the dynamics of the market microstructure. That also raises the question of how do you model in mathematical terms the interaction of many agents who trade and place their orders. To develop mathematical models for the microstructure of financial markets is a very challenging research program which calls for methods developed in the theory of interacting particle systems. I've recently been involved in some related issues in joint work with Ulrich Horst, a former PhD student in Berlin who is now at UBC in Vancouver, and with Alan Kirman, whom I have already mentioned before.

I: It seems that mathematical finance is built on axiomatic and abstract principles (like the efficient market principle). Have these principles been tested and verified?

F: The efficient market hypothesis comes in different forms. In its strong form, it says that the price fluctuation you observe behaves like a martingale. As a special case, it would be the random walk hypothesis, which assumes that price moves like a random walk. What you see is usually not so far from the martingale property, but there is a lot of evidence that you should not take it literally. In fact, that form of the hypothesis is too strong. If you move away from that hypothesis, it means that due to a basic systems theorem of Doob there are strategies that generate a positive expected gain. But there is a weaker form of the hypothesis which is much more flexible. It says the following. There may be strategies with positive expected gain but it is not possible to have positive expected gain and zero downside risk. In other words, there are no free lunches. That makes economic sense because if free lunches were available, there are enough clever people around to seize the opportunity and to wipe them out. In this more flexible form, the hypothesis is widely accepted. There is a broad consensus that you don't find free lunches even though you may be able to make profits with positive expected gain accepting some downside risk. In this weaker form, the hypothesis has been a rich source of interesting mathematical developments. It has been shown that the absence of arbitrage opportunities is mathematically equivalent to the fact that there are equivalent martingale measures. That is an existence theorem. Modern mathematical finance starts on that basis.

I: Can mathematical finance be considered a science?

F: If you translate the question into German, the answer would be clearly "yes". In German, "science" is "Wissenschaft" and "Wissenschaft" is rather broad. It's not just natural science, it also includes the social sciences, economics, and finance as well.

I: If it is a science, one should be able to falsify principles or hypotheses in mathematical finance.

F: Yes. For example, there is a lot of empirical evidence that the efficient market principle in its strong form does not hold. Personally, I do not work under that strong hypothesis. I do work under the weaker one. There is no significant evidence that I know of which would refute it.

I: Does it mean that in finance there are such things as laws that govern the behavior of stock markets?

F: I do believe in the relevance of probabilistic laws in finance. It's reasonable to describe price fluctuations in terms of probability measures on certain path spaces. The absence of free lunches amounts to the existence of an equivalent martingale measure, and this implies that continuous price fluctuations are nowhere differentiable. This can be viewed as a law which explains the erratic price behavior of a liquid stock which you actually see on a mesoscopic time scale. If you take the problem of pricing financial derivatives, you can show that a price must satisfy certain bounds if it does not create arbitrage opportunities. Such arbitrage bounds can be seen as a law, too. Mathematical finance is certainly a science, by my understanding of science.

I: Have you done consultation work for any financial organization?

F: No, I have not done that personally. But some of my co-authors have been involved in that. I have former students who are involved in that. I am following some of their activities, but I try not to get involved myself.

I: It's very lucrative.

F: It may be lucrative, but it also may change your life. I had occasion to watch while working on a joint paper how my co-author was, every once in a while, called to the phone because the program had to be urgently modified in some bank where his ideas were being implemented. I would not like that kind of pressure.

I: What is your advice to graduate students who are keen on a career in mathematical finance?

F: My advice to my own students is to get a broad and solid education in mathematics and not to specialize too early. Even if you decide to work in this area of finance, either in academia or the financial industry, it's a field that evolves rather fast. You need a lot of flexibility, also on a mathematical level. It's not clear that it will be enough to know the tools, for example, needed to understand the Black–Scholes formula. Other challenges may come up which may require very different techniques. I already gave you one example — the microstructure of financial markets. You have to be a good probabilist to react efficiently and to use other methods as well. To my own students, I recommend them not to narrow down too early but to make sure that they are comfortable with a wide range of techniques in probability and analysis.

Avner

Mathematician in Control

Avner Friedman has made important contributions, both in theory and applications, to partial differential equations, stochastic differential equations and control theory. His career, especially during the past two decades, epitomizes a personal mission and relentless drive in bringing the tools of modern analysis to bear in the service of industry and science.

His distinguished career began at the Hebrew University, Israel and weaved, in a somewhat colorful way, through Kansas, Indiana, Berkeley, Minnesota, Stanford, Northwestern and Purdue, culminating in the directorship of the Institute for Mathematics and its Applications (IMA), Minnesota (1987–1997), Minnesota Center for Industrial Mathematics (MCIM) (1994–2002) and Mathematical Biosciences Institute (MBI) of the Ohio State University (2001–). He is also the Distinguished Professor of Mathematical and Physical Sciences at Ohio State University, the latest in a chain of numerous distinguished professorships in the universities he has passed through.

Friedman

His service on many US national boards and advisory committees is an indication of his boundless energy and selfless efforts in promoting the applications of mathematics and advancing the mathematical sciences. Among the honors and awards he received for his wide-ranging contributions are the Stampacchia Prize, NSF Special Creativity Award, and membership of American Academy of Arts and Sciences and of the National Academy of Sciences, USA. He has served and continues to serve on the editorial boards of numerous leading journals in analysis, applied mathematics and mathematical physics. His prolific research and scholarly output has resulted in more than 400 publications, written singly and jointly, and 20 books. He has always been in demand for invited lectures in and outside the US even at the biblical age of three score and ten and beyond, he is directing a concerted effort to bring problems of the biosciences within the reach of the mathematical sciences.

As a founding member of the Scientific Advisory Board (SAB) of IMS since 2000, Friedman has contributed to the development and success of the Institute in its first five years. On his annual visit to the Institute, he was interviewed on 6 January 2006. In the following edited and vetted account of the interview, one can feel the palpable excitement of applying mathematics to the real world and of being drawn into the personal world of a creative and gregarious personality.

Imprints: What was the topic of your PhD thesis? Did it set the general direction of your future research?

Avner Friedman: My thesis was in partial differential equations. It dealt with several different subjects. I have been involved in differential equations my entire career, but have also diversified to other areas.

I: You didn't change fields?

F: I didn't change fields in the sense of going from partial differential equations to algebra. But within partial differential equations, I diversified to a number of areas. Partial differential equations are used in, for example, control theory, applications to industry and, recently, mathematical biology.

I: You went to University of Kansas immediately after your doctorate. Was there any specific reason for this decision?

F: One chapter in my thesis dealt with the so-called problem of unique continuation. Professor Nachman Aronszjan, at the University of Kansas, had done some very important work on unique continuation. I wrote him about my results, and, soon afterward, he invited me to come as a research associate to his department. I was there for one year.

I: From your publications, it seems that initially you were primarily interested in the theoretical aspects (analysis) of partial differential equations but very soon afterwards, you also did and continue to do a lot of work in applied areas like control theory and stochastic differential equations. When and how did that happen?

F: I have always worked on partial differential equations, and I have looked for areas where they can be applied. That's why I started to migrate into areas of applications such as control theory. For a while, I went completely into stochastic differential equations because there were interesting problems in game theory, that is, stochastic games: problems of pursuit of objects when only partial information is known. It turned out that this topic was very well connected with partial differential equations, and I came back to doing partial differential equations through stochastic differential equations. By exploring these applications, I enriched my areas of knowledge and research.

I: Did those applied problems contribute new insights or new developments in partial differential equations?

F: Absolutely. They were very exciting problems. I started to be interested in real applications in the late 1980s when I was exposed to problems in industry that some of my colleagues, especially in England, were tackling. Later on, I moved from Northwestern to Purdue and then to Minnesota to be the Director of IMA (Institute for Mathematics and its Applications). By that time I was completely immersed in problems from industry, and I found out that a large number of theoretical problems in partial differential equations came out of industrial problems.

I: From your large number of publications, it seems that not only are you prolific in writing papers on your own but you also enjoy collaborating with a lot of people. How much of this is due to your own personal temperament and how much to a research philosophy that is consciously pursued?

F: I think that if you look at the trend in mathematics, you will see that increasing numbers of papers are co-authored by two, sometimes three people. More and more, mathematicians and mathematical scientists are talking among themselves. It is extremely stimulating to do so, especially in applied areas. Many of my first papers were done alone, but most of my work now is joint. I often collaborate with others, particularly my former students.

I: Unlike the term "applied mathematics", the term "industrial mathematics" is a relatively new one. Could you tell us briefly what exactly is "industrial mathematics"?

F: In applied mathematics, you pick up problems from the sciences, engineering and other academic disciplines; you may look at the literature to find out where the problems are and try to solve them using mathematical tools. You may discover new mathematics. In industrial mathematics, by contrast, you go to industry to find the problems. The problems are not usually published, and you have to talk to people. You have to find out what those in industry are interested in *today*, because tomorrow they may be interested in something else — or they may be out of job. Find out what they are doing now, what is interesting to them and what the time horizon is for solving the problems. Then you may talk to them, or to your colleagues, or simply think by yourself to come up with suggestions for a solution. You don't necessarily need to find complete solutions. If you publish a paper in mathematics, you must present complete proofs. In industrial mathematics, you may get a 90% instead of 100% solution, but you must get it in a timely fashion.

I: Is work in industrial mathematics usually acceptable to journals in mathematics for publication?

F: Oh, yes. In the IMA, we had a seminar for industrial mathematics, and we had about 25 speakers every year coming from industry. Each one came with a different set of problems. I wrote up, and sometimes rewrote, the problems. There are 10 volumes of these, each containing about 25 sets of problems in a particular subject. About 50 publications were based directly on these problems. There are another 50 papers that might be called second-generation. For example, there was a lot of work done in optics, in scattering, that came from my contacts with Honeywell and some other companies. This has been pursued by some of the people at the IMA, some of my students and postdocs, and they are still working on them with Maxwell's equations. There is a stream of papers that has come out of industry.

I: You mentioned that you published a series of volumes on industrial mathematics. They are not papers but actually books.

F: Yes, in each chapter, there is an introduction to the industrial problem, and then I formulate open problems for mathematicians.

I: It's quite encyclopedic in scope, isn't it? This must have required tremendous energy.

F: Yes, but energy is a function of enthusiasm.

I: Has any of your applied research been used in industry?

F: Absolutely. Work that we have done in optics, called "diffractive grading", was used by Honeywell in order to get grants from the defense department. Also in collaboration with postdocs, I did some work that led to patents at Ford Motor Company. Work that I did myself involving semiconductors and modeling was used by Motorola in chip design for instrumental control to control the acceleration of a car.

I: You are also working on problems in biology?

F: Well, that's what we do at the MBI. I am personally fascinated by the mathematics of cancer, which happens once again to involve partial differential equations.

I: What is your most satisfying piece of applied research work?

F: Well, I think the most satisfying piece of research is whatever I'm working on now. Whenever you work on a problem, it is the most exciting thing in your life for the time you are working on it. If you work in a field that is rapidly developing, it's not just one paper but a sequence of papers. Right now, we have a very interesting line of research that is motivated by cancer, but is nonetheless pure partial differential equations. This is the question of bifurcation problems in free boundary problems. The solid tumor is a moving region, and you don't know how it's going to move and grow. It develops fingering and so on. We try to prove theorems for moving boundaries with fingers, developing fingers as bifurcations. This is really an open area of problems.

I: Do you have to talk to other people like biologists?

F: I would say that I *get* to talk to biologists, specifically to experimentalists who work as oncologists. Of course, I also talk to other mathematicians working in partial differential equations.

I: Do the biologists seek you out to solve their problems?

F: At first, I go to them. When they are convinced that we are actually useful to them, then they also come to us. That has been my experience.

I: But it's not very easy to convince a pure mathematician to go and solve those problems.

F: It's not easy at first, because you have to do a lot of work before you can be useful to the biologists. You have to learn a lot. But I started the MBI because I was certain that mathematicians could make key contributions in the biosciences. Now, it's my personal research interest and my administrative role combined. And we have 14 postdocs involved in different fields of the biosciences. Some of them work on cancer and others on neuroscience, physiology, ecology, genomics, etc.

I: Can you tell us something about the IMA and MBI?

F: The IMA was started in 1982. There was a national competition for mathematical institutes. The NSF decided to have two — one in Berkeley in core mathematics, and the other in Minnesota in more applied work. Hans Weinberger was the director of the Minnesota institute for its first five years. I succeeded him as director. At that time I started to emphasize interaction with industry in addition to general applied mathematics. My point of view was that applied mathematics could only gain wide acceptance, say, in industry, if those doing mathematical research in industry knew you actually could connect with and care about the problems with which they were dealing. In addition, I thought you would find very interesting problems in industry, so I started to visit companies. Typically, I would spend two days in one company and talk to about 20 people. Out of these, I would identify one or two people whose problems might benefit from mathematical input. I would then invite them to talk in my seminar.

After 10 years, I stepped down from IMA, and started the Center for Industrial Mathematics in University of Minnesota. It is a degree program. Graduate students who want degrees in applied and industrial mathematics spend a summer internship in a company and come back to author a master's thesis. Some of them continue to write PhD theses supported by industry.

When NSF called for new proposals, I was already interested in the opportunities biology was bringing to mathematics. I was in Minnesota at that time, and you can't expect NSF to support two institutes in one department, so I worked together with people at Ohio State University to write a proposal in mathematical biosciences. It was a good time for OSU: the medical school was hiring many new people in biological sciences, and people in statistics were very active in biology. Our proposal was successful, and I became the first Director of the MBI.

I: What is generally understood as "applied mathematics" in the United States?

F: Keith Moffatt and I have talked about the fact the "English applied mathematics" has a different flavor from "US applied mathematics". To give you a flavor of US applied mathematics, materials science is an important area of applied math in the United States. You can use mathematics in the modeling of it. For example, car companies want to increase mileage per gallon — it's a government requirement. To do so, they want to replace steel with lighter material, say aluminum. But aluminum is not strong enough, so they add carbon particles to make it stronger, and it turns out that partial differential equations can be used to predict how this new material will behave. Ford Motor Company actually came up with a problem and we did some work on it at the IMA.

I: Are these predictions successful mathematically?

F: Yes, it turns out that the predictions have been very useful to the engineers. As a result, the field has completely changed since our first materials science program in 1985. The mathematical community of people working in materials science has increased tremendously. Other examples of US applied mathematics come from applications in control theory, computational science, applied linear algebra, fluid dynamics, scattering theory, nonlinear waves in oceans and materials, polymeric materials and polymers.

I: What about operations research?

F: Operations research applications have ranged from manufacturing to finance, and there is so much more. Imaging has developed rapidly in many aspects: imaging distant targets is a different problem than imaging at the molecular level. Speech recognition — we have a volume at the IMA in speech recognition — involves Markov processes. Applications even come in from traditionally pure mathematics. The field of US applied mathematics is vast and diverse. We had programs in applied number theory, in coding, communications, graph theory, scientific computation as well as fluid dynamics.

In England, by contrast, fluid dynamics used to be the crowning theme, because England is surrounded by water. Traditionally, England is very strong in computational fluid dynamics, and they are looking at all kinds of phenomena in waves and fluids. Many of the mathematicians working on these problems inspired me to get involved in applied mathematics in the first place and ultimately to bring industry to the table to expand the kinds of problems mathematicians are involved in solving.

Roe

Mathematics, Music, Masters

Roe Goodman of Rutgers University was interviewed on 11 February 2003 at the Department of Mathematics, National University of Singapore while he was visiting IMS and the Department of Mathematics. He was a guest participant in the IMS program on *Representation Theory of Lie Groups*. The hour-long interview covered topics that range from teaching and research in mathematics to the influence of masters in mathematics and music.

Goodman's extensive research activities are centered around Lie groups. Together with Nolan Wallach, he has written a 685-page encyclopedic book *Representations and Invariants of the Classical Groups* that is both an introduction to as well as an authoritative reference on the structure and finite-dimensional representation theory of the complex classical groups.

Goodman

Imprints: Can you share with us some of the excitements of your latest research?

Roe Goodman: My own research started in the 1960s when I did my PhD thesis with Irving Segal in MIT. Segal himself was primarily interested in the mathematical problems of quantum field theory, viewed in a very broad sense: difficult questions in nonlinear partial differential equations and their symmetry groups. My own interests and activities over the years have moved in the direction of representation theory and the symmetry groups although I maintain an interest in application to physics. The thing that attracted me to representation theory is that it lies at the crossroads of all of mathematics. You have the analysis side in connection with partial differential equations and you have algebra and geometry in the Lie groups. One of the things that is exciting about this field, as I have watched it developing over the last 40 years, is to see so many areas of mathematics come into this field — more and more of combinatorics, geometry and algebra — even though the subject started out with a lot of emphasis on analysis. Of course, one of the central things that make mathematics research so exciting is that, over the course of time, you see that problems that first seemed intractable examined by lots of people who find new ways to approach these problems. For example, problems that were originally posed as questions of functional analysis can now be approached using algebraic techniques, which simply avoid some of the difficult, maybe impossible, analytical problems. I have spent quite a lot of time over the last decade telling the story for the next generation, so to speak, in my collaboration with Nolan Wallach. We wrote quite a big book on representation and invariant theory, trying to make the basic results and philosophy of representation theory accessible to the current generation of mathematicians (and we hope to another generation).

I: Are there any unifying trends in the development of your field of research? Do you think that particular problems have to be solved first before some unifying theory can arise, or do you think that essentially new theories and concepts rather techniques need to be proposed before outstanding problems can be resolved?

G: In my field, it seems to be that there is this cycle of solving particular cases and pushing the methods that suffice for those cases as far as possible. At a certain point those methods often turn out to be insufficient or the computational difficulties simply become insurmountable and then there are new approaches that come in. One of the striking things about mathematics is the insistence to understand the subject from the conceptual point of view. For beginners of the subject, it is hard to understand the concepts without actually doing some calculations. But at a certain point, you discover that even if you have a very powerful computer doing symbolic calculations for you, the calculations alone are not going to tell you what the pattern is. You have to discern the pattern, and I think finding the pattern is one of the main purposes of mathematics. Of course being able to come out with an answer that can be translated into some of the applied domains is also very nice when you can get it.

I: In physics there is some kind of blueprint for the development of the subject whereas in mathematics there is no specific blueprint as to how mathematics should develop.

G: That's right. The remarkable thing in mathematics is that you have these extraordinary imaginative people (like Gromov, Langlands) who propose concepts that, to ordinary mortals like us, seem to just come out of the blue. Of course, they have a basis for those ideas but it can take the work of a lot of people to develop the consequences.

I: There is some perception that pure mathematicians look upon practical applications with disdain. How much of this is true? Some of the best mathematical minds like Hilbert and Poincaré have worked in both pure and applied areas. Is it possible to achieve their status in the present age of specialization?

G: Judging by my own experience, it seems hard to establish links with applied science departments like chemistry, computer science and physics partly because the faculties in those departments themselves have quite a high mathematical level and they generally view the kind of mathematics they are using as something that they are reasonably competent with and they don't seem to have an enormous need for mathematicians. Of course, one can try to create the need, and there is also the tendency on the mathematician's side to think, as the phrase goes, "We would rather build fire houses than to put out fires". But there are remarkable counter-examples. My own personal hero is Hermann Weyl, who is not very well known to the general public. He was a student of Hilbert. He gave the first set of lectures on Einstein's general relativity theory in 1917/1918 and published basically the first book on general relativity theory based on those lectures. He also worked as a forerunner in understanding and explaining what was going on in the new quantum theory in the 1920s. Certainly there are many examples of mathematicians who have done this. In recent years someone like Irving Segal is an inspiring example. Another person who comes to mind is Michael Atiyah. As physics becomes more mathematical and uses a wider range of mathematical tools like algebraic geometry, physicists have to turn into mathematicians.

From the point of view of people completely outside of science, and in particular people interested in what is the worth of mathematics to society at large, they would like to see how mathematics can be turned towards more practical things. It is interesting to observe that in my own field involving harmonic analysis as well as representation theory very recent work in wavelet analysis is essential for things like image compression and data analysis. A lot of that grew out of what used to be thought of as quite abstract kind of harmonic analysis and abstract Fourier analysis. It is a question of having links with applied mathematics. There are people like Coifman at Yale, who, when I first knew him in the 1970s, was working in the kind of representation theory and harmonic analysis that I was. He moved into wavelets and has been very successful in promoting its commercial technology.

I: What instrument do you play?

G: The instrument that I play most seriously is the bassoon. Originally I started out as a child playing the cello but then when I was 14, I switched to the bassoon, which I consider the most nonlinear oscillating system that is of any practical use. So every morning when I practice my instrument I perform experiments on a little nonlinear oscillator in the form of a bassoon reed.

I: What made you switch from music to mathematics?

G: My father was a professional musician, a pianist, and I knew from personal experience the difficulty of making a living as a musician. Most of my adult friends in the orchestra that I played in as a teenager advised me that it would be much better to go into science. But it was hard for me not to go into music and composition because that was what I was most passionately interested in at the time. In retrospect, however, mathematics has been a very rewarding career, and I have still managed to maintain an active musical life.

I: Do you think that part of the problem of overspecialization is a lack of communication between mathematicians and people in other areas like engineering, physics and computer science?

G: Yes, that is certainly a problem and I think that is a real challenge for mathematicians. One can almost feel it as a drawback of mathematics that we have such a perfect system of notation that for us the notation serves all the purpose that we want in the same way as the written language serves our purposes. But students and people outside of mathematics often tend to view mathematics simply as a collection of symbols to push around. When the symbols get too complicated, only the professional mathematicians can read them and then people outside the field just turn off. I don't know how to get around that. I teach engineering students a lot and try to explain the concepts in a way that is acceptable to them. I view that as one of the biggest challenge when I am teaching. Of course, it can be quite frustrating because you know as a professional mathematician that with the benefit of an appropriate concept certain ideas can be quite simple. But this is only true if the person dealing with the concept has mastered it, and for people outside mathematics the notation and concept can be so obscure that it is very hard to get the ideas across. I think that is one area in which mathematics, as a profession, sometimes tends to be too narrow. We don't realize that the mathematical ideas are just too dry when used by students outside of mathematics.

I: There are some people who would attack a problem from first principles. They develop their own understanding of the problem and then develop essentially their own methods for the problem.

G: I think the most spectacular example in my own field is Harish-Chandra who, starting in the late 1940s, simply came into the subject of representations of semi-simple Lie groups on his own. There had been very important preliminary work by the Russian school under Gelfand, but Harish-Chandra started at the beginning and created an incredible edifice single-handedly. For a period of about 25 years, starting from the late 1940s to the mid 1970s, he was so clearly leading the field. that it was only in the early 1970s that there was a significant number of other people working in the field. In his case, the methods were always his own. He took what were, in some way, very classical methods and extended them to serve his needs. It has taken several mathematical generations to go beyond Harish-Chandra's methods. His ideas had tremendous depth. Of course, now more recent approaches to the subject try to understand it by other methods, but he basically set the direction in the field. The results achieved were so precise and profound that everybody in the field has to take his methods into account. A parallel instance in mathematics of someone creating a monumental edifice is in algebraic geometry. Grothendieck created very general machinery that has now become the language of algebraic geometry. So I think the absolutely strongest people in the field simply create the field by using their own methods and then the rest of us have to learn those methods and see what other results can be obtained.

I: Is there a role for perseverance? How much inspiration does one need?

G: Oh, absolutely. I think without perseverance you certainly can't do mathematics. If there are never any ideas that come along, it is pretty discouraging. It is an elusive thing. Solving a mathematical problem is trying to judge at any moment whether the track that you are trying is going to pan out. Of course, perseverance alone may not work, but even if it does, you try to know whether you are moving towards a dead end. That can be very discouraging in mathematics.

I: Do you think that mathematics is a marathon race that is long, arduous and lonely?

G: I think there is a partial truth in that comment. But there is such a large social element in mathematics, public perception notwithstanding, in the sense that if you only create mathematics in writing and never tell anyone about it, then it is like running a long race where no one is even looking. I like to think that at least there is this aspect of mathematics as a communal effort. As Einstein commented, there are innumerable problems in mathematics. But I think the ones that have a life of their own are the ones that have a significant number of people (which, of course, in mathematics could be a small number) with some real interest in those problems. And then the joint efforts of people working on these problems make it interesting — you get some results yourself and compare yours with what other people have. So maybe instead of thinking that it is a long marathon race, it is more like a situation I observed once, to my surprise, at a rehearsal of the orchestra. A grand piano was on the floor of the concert hall but needed to be on the stage. I certainly couldn't lift it by myself, but with eight people it was very easy to lift the piano onto the stage. So I think hard mathematics problems may have some of that element of joint effort. Of course, it is one thing to get the piano onto the stage and another thing to get a beautiful performance. We do need the gifted mathematician to give the beautiful performance but the joint effort can play an essential role.

I: Do you think we have now reached the golden age of mathematics? Do you think it is largely due to the easy and swift access to and dissemination of information provided by information technology?

G: I certainly think that the possibility of going online and trying to get access to a large number of journal articles without having to leave your office or home or wherever in the world you are is quite wonderful. I know how different it used to be when I went to New Hampshire for the summer and carried a lot of journal reprints with me. On the other hand, I think that information technology or even a powerful computer is not enough to create mathematics, although it can give clues as to what is possible. Mathematics is basically created by mathematicians and not by the technology. The technology certainly plays an important role in communicating mathematics because now it is so incredibly much easier to turn mathematical ideas into printed form. This enables us to communicate ideas and to put mathematical papers in online archives so that people know right away when something new appears.

Bryan T.

Viral Visitations, Epidemic Models

Bryan T. Grenfell made important contributions to population dynamics with his pioneering work on the mathematical modeling of infectious diseases like measles and whooping cough, foot and mouth disease in farm animals and influenza of avian, equine and human types. He has done extensive collaborative multidisciplinary work at the interface between theoretical models and empirical data in population biology.

He has worked at York University, Imperial College, Sheffield University and Cambridge University, where he was Professor of Population Biology, before moving to the Center for Infectious Disease Dynamics at Pennsylvania State University in 2004 to become the Alumni Professor of Biology. He has played advisory roles to the British government on the foot and mouth epidemic in 2001 and to the National Institute of Health (United States) since 2002. He is also active in organizational work of scientific meetings. He has served on editorial boards of leading journals in theoretical biology and ecology, and currently of *Public Library of Science Biology*.

His scientific contributions have earned him the T. H. Huxley Medal, Scientific Medal of the Zoological Society of London and Fellowship of the Royal Society. He was also awarded the Order of the British Empire for his services to epidemiology and the control of infectious diseases.

Grenfell

He was Chair of the Institute's program (August–October 2005) on the mathematical modeling of infectious diseases and was interviewed on 24 August 2005. The following is an edited and enhanced version of the transcript of the interview, in which he traced his transition from traditional zoology to his pioneering modeling work in population dynamics on infectious diseases. Here he gives us an insight into the multidisciplinary richness of a fast-growing area that is not only of immediate importance and urgency but also intellectually challenging.

Imprints: Was your original training in zoology a traditional one? How did you get into your present research interest?

Bryan T. Grenfell: My training was indeed a traditional one. It was a zoology degree in Imperial College, London. I wasn't a great field zoologist or person in the lab, so when we then had a course in the final year on population dynamics in ecology, I seized on it with open arms. I then did a PhD on applied ecology, specifically the application of models and statistics to assessing whale population sizes in the Southern Ocean. For my first postdoc, I worked, again at Imperial, on parasitic worms and childhood infections. I then got a faculty job in Sheffield University and I've worked on infectious diseases since. So the disease theme is since the 1980s.

I: Practically from the beginning, you were already quite theoretical.

G: Yes, reasonably, though I'm a biologist, not a statistician or a mathematician.

I: How old or recent is your field of research?

G: It really goes back to Daniel Bernoulli in the 1700s and then a body of work in the 19th and 20th centuries on infections like small pox, malaria and measles. The importance of these infections in public health terms and the quality of the data and simplicity of some of the mathematical patterns led people to use statistical and mathematical modeling approaches. I guess we can think of giant figures like Ross, Bartlett, in the 20th century. And in the late 20th century, the importance of infectious diseases that we all know about has led to another explosion in applications of mathematics and statistics in this area. There's been an explosion in disease dynamics work since the late 1970s, catalyzed by the seminal research of Anderson and May. I think it is a very lively field and it melds basic questions and applied questions in fields all the way from mathematics and statistics to immunology, virology, population dynamics and evolutionary biology. So I think it is a very exciting field.

I: So it's really quite an old field.

G: Yes, older than many others in terms of applications of population dynamics.

I: In your modeling work on infectious diseases, which came first — the empirical data or the theoretical model? In other words, do you look at the data and then formulate the model, or do you first form an intuitive model with which to compare the data and then subsequently refine it?

G: I think the overall answer is that one does both: the model and the data should be very closely linked and co-evolve. A lot of these biological systems, particularly the ones which manifest themselves at the level of interactions between people, are very complex, potentially with many parameters. The more we can tie up by comparing models with data the better. We are lucky that, because of historical notifications for many important diseases, there are sometimes very good data; for measles for example. But I must admit that I often go into the preliminary statistical analysis and so on with an intuitive model, then build a more formal structure.

I: Where do you get the data from?

G: A lot of infections were notifiable; they had to be notified: measles and pertussis (which is whooping cough) in the UK, for example. Today, such incidence data are supplemented by freely available data on the genomic variation of influenza and other viruses. The explosion in molecular genomic data is very exciting.

I: Are new statistical techniques needed to deal with the large data sets that you are faced with?

G: Definitely; I guess there are there are three parts to the answer. Focusing on our work, we use wavelet analysis to explore highly nonstationary epidemic dynamics in the frequency domain. We then use mechanistic nonlinear autorepressive models to estimate epidemiological parameters. Your colleague Yingcun Xia has done seminal work here. Finally, we are now trying to unify these population dynamic analyses with phylogenetic approaches to viral molecular data.

I: It sounds very cross-disciplinary.

G: Very, because it blends statistics and mathematics with epidemiology, virology, immunology and evolutionary biology.

I: Did you have to pick up the mathematical ideas and techniques on your own?

G: Yes, though I've also been very lucky with wonderful technical collaboration.

I: Is your field connected with evolutionary biology?

G: Yes. Originally a lot of my work was on straight population dynamics. Once you get into influenza, you've got to think about evolution. I'm increasingly getting interested in that.

I: Is biology getting more mathematical and statistical?

G: I hope so. After the genomics revolution, biologists are very interested in systems biology now, which is the interaction between genes and their products leading to gene regulatory networks. If you have huge networks, you have to have some theory. So I think that the laboratory people are now using dynamic approaches.

I: Can computer simulation models be used for predictive purposes?

G: Let's imagine the case where you make these models before an epidemic. I certainly think they are very useful for projective purposes. They are very useful for saying, based on our assumptions about how people mix and the characteristics of the disease, what would we project would happen under different control scenarios. I certainly think that it's great to have such complex computer models but you also must have simple models — more reductionist models, just so you can interpret things. Quantitative prediction before a disease has hit is very hard; however, simpler "operational" models, fitted to the early part of an epidemic can be useful.

I: Do I understand that they have actually been applied in actual projections?

G: Yes; for example, we and others worked on the foot and mouth disease epidemic in the UK in 2001. A family of models was used with a range of level of complexity to project what was going to happen and they tended to be useful in making qualitative inferences about what sort of control policies one might have to adopt. Having a range of models which all pointed in the same direction was useful for the policy makers here.

I: Are there any specific models being formulated for the recent SARS outbreak?

G: I think if you look at the literature, there's a range of models that have been made by many groups, particularly using the high quality data from the major outbreaks.

I: Has any work been carried out to determine whether epidemics play a role in the evolutionary history of birds and animals?

G: Oh, certainly. Again it's not central to my work, but a lot of research has been done by geneticists, for example, looking at how some relatively stable parasites like herpes viruses co-speciate with their hosts, but also the impact of malaria, for example, and its interaction with human genetics. Then, more fundamentally, there's a lot of work on the possible role of parasitism in the evolution of sex, for example.

I: Are most epidemics in human history the result of human actions?

G: Not in any simple sense, no; though colonization, anthropogenic changes like deforestation can play a role.

I: Are there any models for cross-species pathogenic evolution?

G: Not so much models, but I think the biologists are getting closer and closer to understanding the species barrier — why does a virus grow in one species and not in another — sexually transmitted is all we care about. The flu virologists are getting closer to understanding what those barriers are, and that's true for a variety of other viruses as well. But there are still always going to be the big questions to answer, particularly for more complex parasites.

I: Are there any past records, from paleontology, say, or something, to show that epidemics could have wiped out a whole species?

G: I can't think out of the top of my head that there are certain cases like that. What you might expect is that if it is a self-sustaining epidemic in a population of hosts, the epidemic often drops out before the hosts do. But if you have a big population of one species and a small population of another living cheek by jowl with it, and then you have a species jump from the big species and which could continue to jump across, you can then imagine that the small population would be very endangered by the disease. African wild dogs' diseases are certainly a problem in small populations. [The same goes for] gorillas in Rwanda and measles and so on. In small populations, of course, there's always a danger that the disease will just exert back extra toll and wipe the infection out. But I don't know of any example — there might well be one in history of a big population that's been wiped out by its own diseases. Because the disease co-evolves, it often becomes less pathogenic.

I: Are there any models that predict the onset of resistance to certain diseases say in an epidemic?

G: You mean things like antibiotic resistance? There are certainly models that people have used to try and understand how the evolution of antibiotic resistance is facilitated by how hospitals are managed or how the development of resistance against drugs which control parasites in farm animals, for example, depend on how the drug is used. Often though, it's direct statistical experimental evidence that's needed there for such models.

I: How much of the models are related to dynamical systems?

G: Pretty much all of it. For example, measles is a classical example of a (seasonally) forced oscillator. However, as we add more biology, things become more complex. For instance, measles can go [through] extinction epidemic troughs, implying a discrete state space system. As another complex, spatial heterogeneity and network mixing are often important. However, the very simplest models can still give insights.

I: Do you do consultation work for the government and others?

G: For the foot and mouth epidemic, I was a member of one of the modeling groups that advised the government. I also do some advisory work for WHO.

I: Do you have many students?

G: In Cambridge, I had a big group. There were 15 or 16 of us — maybe eight or nine postdocs and the rest graduate students. Having moved to the US, I'm building up the group again now.

I: What you do is very critical for health control in populations...

G: It's certainly got a strong applied aspect, but all the people in this field also do it because the questions are very interesting. I love dynamic processes and spatial processes, and the epidemiology is very interesting in that way.

I: Do you have any advice for people who want to study these things?

G: I certainly think it is a growing field and will grow much more over the next decade. There are great problems, wonderful data and great opportunities for people, particularly people with the right technical training in statistics or physics or mathematics. I know a lot of brilliant young people who have jumped across from these fields.

I: Will it be easy for a mathematician who knows nothing about biology to cross over?

G: Certainly I have several people that I can think of who've done it brilliantly. Several people — again, Dr Xia for example knew no biology and did some wonderful work. Another postdoc did his PhD in astrophysics on galaxy simulation and now has a faculty job doing epidemic modeling. But, as these people did, you have to be prepared to learn and to realize that biology is complicated and that a key thing is to really get in amongst the data.

I: The mode of thinking in biology is very different...

G: It certainly is. A lot of the ideas are qualitative and you have to respect the fact that the folks who have been in the field or lab for a long time have got a sophisticated model understanding what is going on. As more and more data are collected on dynamic processes, quantitative skills are really important to interpret them. So I think it's a great field to get into. I encourage people to do that.

Takeyuki

Brownian Motion, White Noise

Takeyuki Hida (飛田武幸) is well-known for his pioneering work in establishing and developing a new field in probability theory — the field of white noise analysis, which has now found numerous applications outside probability in quantum dynamics and biology, and within mathematics itself in differential equations and geometry.

His doctoral thesis sowed the seeds of a new type of differential and integral calculus (now called the Hida calculus) for Gaussian processes in terms of the time derivative of Brownian motion. This was developed further in his analysis of generalized white noise functionals, first proposed in his Carleton University lectures in 1975. In the decades that followed, he initiated a program of investigations into functionals of general noise, in particular Poisson noise, and the analysis of random complex systems. He has also applied his results to provide new approaches to Feynman (Lagrangian) path integrals and the Chern–Simons action integral and to problems in mathematical biology.

Born in Okazaki in Aichi Prefecture, Hida had his undergraduate education at Nagoya University and obtained his PhD from Kyoto University under the official supervision of the distinguished probabilist Kiyosi Itô, the founder of stochastic analysis. Immediately after he obtained his BSc, he taught for seven years in a teachers' college, Aichi Gakugei University. Subsequently, he taught at Kyoto University (Yoshida College) for five years before joining Nagoya University

Hida

as Professor of Mathematics. On his official retirement in 1991, he was bestowed the title of Professor Emeritus by Nagoya University and he also took up a professorship at Meijo University. Since 2000, he holds a Special Professorship position at Meijo.

At an age when others would be content to bask in past achievements, he continues to collaborate with an active team of researchers in Nagoya University in pushing the frontiers of his discipline further afield into the scientific unknown. Hida's research output includes more than 130 research papers and seven books. He has been invited to leading universities and major scientific meetings in the West. He has served as Dean of Science at Nagoya University and as Dean of Science and Technology at Meijo University. He was Chairman of the Committee of Conference, Stochastic Processes and Applications. For his scientific contributions, he was awarded the Chunichi Cultural Prize in 1980 and made an Aoi Citizen of the city of Okazaki. More recently, in 2007, he was awarded the Zuihou Jyuukou Shoh (瑞宝 重光 章), one of the highly prestigious awards in Japan.

Hida's connections with NUS go back to 1981 when he was an invited speaker at the International Mathematical Conference organized by the Department of Mathematics, NUS. Since then he has maintained close ties with mathematicians in NUS through official invitations and personal visits to Singapore. He was invited by the Institute and the Department of Mathematics to give colloquium lectures and seminars on white noise analysis in April 2007. He was interviewed on 2 April 2007 during this visit.

The following is an edited and enhanced version of the transcript of the interview in which he traces the emergence from research isolation in a teachers' college to international prominence in the world of probability and leads us through the excitement of a newly emerging field that is as profoundly abstract as it is diversely applicable.

Imprints: You had your undergraduate education at Nagoya University shortly after the war at a comparatively late age. How did it affect your studies?

Takeyuki Hida: Not quite at a late age because the educational system in Japan at that time was different from the present. Usually you graduate at 23, starting from elementary school to junior high school, high school and university. In my case, 24 — so not much difference. However, compared to other people, I was delayed for almost two years.

I: You taught at Aichi Gakugei University for seven years before going to Kyoto University where you got your PhD. Was there any particular reason for this comparatively long time gap?

H: Of course, I had wished to go to graduate school sooner. But when I finished my undergraduate studies, I immediately took up a job. I had to work because I came from a farmer's family and I had brothers and sisters. It was financially difficult for me to continue with my graduate studies. I had a responsibility to my family. Aichi Gakugei University is actually a teachers' college. I only had to teach and I had heavy teaching duties. I didn't have much time to do research. I corresponded with Professor Itô, of course. At that time, he had already left for Kyoto University but I was able to visit him sometimes. In addition, I communicated with Professor Paul Lévy who was in France. The reason was that as an undergraduate I was taught by Professor Itô, and the textbook used was Lévy's book [*Processus stochastiques et movement brownien*], starting from Chapter 6 on Brownian motion. So I felt very close to Professor Lévy, and also I wanted to hear from him — his comments and suggestions, etc. I was then teaching at Aichi Gakugei University.

I: Did Professor Itô suggest a topic for your research?

H: Not in particular. In my last undergraduate year, he suggested that I read about Brownian motion from the book by Paul Lévy.

I: Your PhD topic was on Gaussian processes. Why did you choose to work on this topic? Did Kiyoshi Itô have any influence on your choice?

H: He did, and so did Professor Lévy. The main reason is that I started studying Brownian motion from Lévy's book. Brownian motion is very important and more basic than the Gaussian process. Professor Itô was my supervisor in my last year as undergraduate. At the same time that I graduated as an undergraduate, he moved to Kyoto. There was no direct communication afterwards, but sometimes I did go to Kyoto, but not often. We usually communicated by mail. With Professor Itô in Kyoto, how could I continue my studies? I had wished to be directed by him. Once I asked him what action I should take. Then he replied that I should follow the pioneers' work — pioneers, he said, like Kolmogorov, Feller, Lévy. He didn't include himself but I add: I understood that a pioneer's work was difficult and not easy to understand, and I found that his [Itô's] paper was interesting to follow. However, he suggested that I should investigate the pioneer's work and find out what the pioneering idea was and that I should do it by myself. I think this is a very nice suggestion.

I: Professor Itô himself also followed the pioneers.

H: I cannot say. I followed his work.

I: Did your PhD work have any influence on your later research?

H: Sure. That was very important. Actually, in 1955, there was a famous paper by Paul Lévy at the Berkeley symposium on the Gaussian process and its close connection with Brownian motion. More precisely, we hope to express any general Gaussian process as a linear function of the Brownian motion — linear meaning some kernel function and some Brownian motion integrated with respect to it. However, Paul Lévy said that the representation is not unique. If there is a way to determine uniqueness, then the combination of kernel function and Brownian motion preserves everything about the Gaussian process. I was very impressed by this idea. I tried to prove the general theory. Unfortunately [it does] not always exist, not even for the single Brownian motion. We may need many Brownian motions that are independent. However, the important part is that the basic process is Brownian motion and the general Gaussian process is expressed as a linear function of Brownian motion. I was thinking about the uniqueness of the representation. In order to characterize the Gaussian process, if there is unique representation, one can say everything about the kernel function and the Brownian motion. For many other representations, it is impossible to say so. There is a very simple example with different kind of representation. If the kernel function is taken to be polynomial, the Gaussian process has more than two different expressions. I was thinking, and the answer was eventually very simple, but the meaning is very deep. That was the idea of the canonical representation. It took me a year to come to a better understanding of this canonical representation. I tried to find out the meaning, I computed an example and finally got the definition of the canonical representation and the uniqueness theorem and characterization. So, the basic part of Gaussian theory is solved.

Given a Gaussian process satisfying some conditions, we can construct a Brownian motion which leads to the canonical representation. The Gaussian process and Brownian motion have the same information — otherwise there is no meaning at all. Given a stochastic process, it would be fine, if we can find some nice basic processes (independent increment process like additive process) such that any process can be expressed as a function of those independent increment processes. Suppose we find the required properties of the process, then we can combine a (nonlinear) kernel function and the basic processes just obtained. The Gaussian case is a very particular case and is, in some sense, elementary. We should generalize. We now come to the general idea. Given a general random complex system, we can find independent (increment) systems and nonlinear functions to express the random complex phenomenon in question. We can think of the general theory of the analysis of those random systems. This is a very big problem. In some sense, it is vague. But still, I wish to extend from the Gaussian process to the general random complex system, including stochastic processes and random fields.

I: Is this a kind of program you are proposing?

H: Yes. This is the original idea of white noise analysis. This is, of course, a very big program. We can do it step by step. First, the basic process is an independent increment process, or I would say, white noise, time-derivative of the Brownian motion. For that purpose, we need some detailed analysis where the functional is nonlinear. I was not dreaming. The actual program was there. That is what I wrote in the Carleton Lecture Notes Series in 1975. Two years ago, we celebrated its 30th anniversary. Many people have referred to the notes. It is included at the beginning of the publication of my selected papers.

I: Has this got to do with your calculus (the Hida calculus)?

H: Yes. That's the origin of it. At the beginning of 1975, at the request of Don Dawson of Carleton University, I gave a series of lectures in the summer, mainly for researchers and some graduate students.

I: You taught at Kyoto University for only five years.

H: There is an unusual story behind this. Because I followed the suggestion of Itô in following a pioneer (Paul Lévy's direction), I communicated with Itô and Lévy when I was in Aichi Gakugei University. Prof Itô suggested to Prof Akizuki, who was then Director of the Department of Mathematics of Kyoto University, to invite me to Kyoto — that was very unusual because Kyoto University is a very prestigious university and I was only an instructor in a teachers' college which was of not so high a standard. I was extremely happy and honored about it, and because of this, my studies very much accelerated.

I: You were in Nagoya University for a long time until you retired. Were you very strongly attached to your prefecture?

H: Nagoya University knew and appreciated what I was doing in Kyoto, and was kind in inviting me as a professor. I consulted with Prof Itô and he agreed that I should move to Nagoya. That was why I moved to Nagoya University. I worked there for 27 years. My hometown is actually Okazaki, not Nagoya.

I: Nagoya University has two parts, one is the School of General Education and the other School of Science, isn't it?

H: Formally, I spent two years in the Mathematics Department of School of Education. Two years later, I formally moved to the School of Science. Nagoya University is very famous in science. There are seven national universities which are very prestigious in Japan — Tokyo, Kyoto, and so on. Nagoya is one of them, and in fact, the youngest.

I: Was there much work done on white noise analysis before 1960 (the beginning of your second research period on white noise analysis)?

H: 1960 was the year of my thesis dealing with Gaussian processes. It is difficult to say [it is] the origin of white noise analysis. White noise theory is, in a sense, a generalization of the study of Gaussian processes. For five years, I studied very hard to understand the meaning of canonical representation using the methods of stochastic analysis. My thesis came from the work of those five years.

I: Could you give us a brief idea of white noise?

H: Take the time derivative of Brownian motion, $\dot{B}(t)$. It is an independent basic system, and the structure is linear. I wish to come to the nonlinear case. That is the first idea. The second step is Paul Lévy's proposal in 1937 in his famous book *Addition des variables aléatoires* (*Addition of Random Variables*) to have integration. There he took a discrete time random process X_n. Suppose we know all the information of X_n until time *n*, then at the next step X_{n+1} is a function of the known value plus independent variable. That was the innovation theory for a discrete time series. Then in 1953, Lévy wrote a booklet published by the University of California, Berkeley, in which he proposed to think of the innovation for a continuous time parameter stochastic process $X(t)$ — he didn't say differential equations — but variation of $X(t)$,

$$\delta X(t) = \phi(X(\tau),\ \tau \leq t,\ Y = \dot{B},\ t,\ dt).$$

To obtain δX, of course, depends on the parameters *t* and *dt* — that are non-random and not of much interest. The important part is that within time *dt*, the stochastic process brings new information, and this new information is independent of the past. He gives a formal expression for the stochastic process, and it serves, in a sense, as a generalization of the classical stochastic differential equation. He asked for a general expression theoretically. I was very much impressed by this idea. So we should find an independent system that will serve to express a given complex random system. The basic variables are independent, the function is a non-random function. So we can combine two mathematical theories. One is a system of independent random variables, and the other is functional analysis. We can then establish a general theory for random systems.

One more motivation comes from Norbert Wiener's approach. In 1940, he wrote about time series and proposed prediction theory. Linear prediction is rather easy but nonlinear prediction is difficult and more important. Sometime later, in 1958, Wiener wrote a famous book on nonlinear problems in random theory. There he discussed many applications of the nonlinear function of the white noise $\dot{B}$. I was inspired, I would say, by this approach. I thought I should study nonlinear functions of $\dot{B}$, not the Brownian motion B, and even more we should establish calculus. So we have to introduce partial derivatives and integration with respect to the variable $\dot{B}$.

In 1967/1968, I was invited to the Mathematics Department of Princeton University by William Feller. He had listened to my Berkeley lecture and agreed very much with what I was thinking and with my plan of research. At Princeton, I was able to give a graduate course and undergraduate course too. Feller appreciated my white noise approach. He had published the third edition of a famous book on probability theory (Volume 1) and at the time I was there. Volume 2 was published a little bit earlier. In the preface of his book, he wrote that he wished to make probability theory a part of pure mathematics. Many people think that probability theory is about gambling and related topics. Feller was not willing to do it in that way. He said that probability theory should be one of the branches of pure mathematics with good connections with analysis. In the weekends, I often met him and he was fascinated by the picture of the chromosome. He was very much interested in applications of probability theory to biology. Two viewpoints are important: one is that it is part of pure mathematics and the other is that we should have good contact with applications to discover good problems in mathematics. The important thing is that though we are applying the theory to applications, it is not quite applied mathematics. We should investigate concrete problems and if we are lucky, we can discover mathematics in the applications. Even though you are studying biology, you are not a biologist — you are a mathematician. We should try to find mathematical theory in biology. That was the way of Feller's research so far as applied mathematics is concerned. I was very much impressed by his idea.

I: Were you surprised by all those connections with conformal groups and geometry?

H: Yes. And it is quite natural. I have not yet obtained good results so far in group theory or good connections with Lie groups. Only for the simple case like $SO(\infty)$. I observed those groups through white noise. The basic part is conformal invariance. That was in Princeton, 1967–1968.

Don Dawson agreed very much with what I did and he invited me to give lectures in Carleton University. Sometime later, I visited Carleton again, in 1975, and he asked me to give a series of lectures in summer, mainly for researchers and graduate students. In this case, I started with a function of $\dot{B}$, starting from Paul Lévy's expression. I had dreamed of a new way to study stochastic differential equations, and I thought it would be fine if we could obtain differential equations in the variable, not x, but in $\dot{B}(t)$. I was inspired by Norbert Wiener's nonlinear networks whose input is white noise, and output is a nonlinear function of white noise. How do you identify the unknown nonlinear network device in between input known and output known too?

I: What are some of the most important applications of white noise analysis?

H: Physical applications — the most important one is the Feynman path integral and related topics in quantum dynamics. Feynman proposed functional integration. Starting from the Lagrangian, he wished to introduce infinite-dimensional integration to obtain the propagator in quantum dynamics. I think we can imagine the original idea in Dirac's book [*Principles of Quantum Mechanics*, 1930]. Of course, Feynman improved it very much so that we can do more. However, the general idea was proposed at that time. It is known that a constant trajectory is determined uniquely by the Lagrangian. In quantum mechanics, the trajectory fluctuates and the fluctuation is expressed as a Brownian bridge — we are suggested that description from Dirac. I have discussed with Streit (German physicist) and he agreed with me. The problem is that the velocity and kinetic energy of the particle are white noise functionals, and we have to perform integration and establish a calculus.

I: What about applications to biology? For example, could one view the "junk DNA" in the human genome as some kind of white noise at the molecular level?

H: A friend of mine called Naka who used to be in New York University Medical Center had applied white noise to identify the action of the retina of catfish, which is simple compared to that of other animals. Naka was clever enough to consider the nonlinear part of the action. He did some very complicated computations, but unfortunately, he passed away last year after returning to Japan. It's a very sad story. Much of his work, however, can be seen in the literature by him and his colleagues.

I: Are there people continuing his research work?

H: I hope so. Another application is in the study of polymers. Biologists study the mechanism underlying what happens when a polymer is cut. They conduct a lot of experiments and computer simulations. I have a good friend in Kyoto who proposed some principles and compared it with actual phenomena. Another friend of mine (Oosawa) is studying the movement of paramecium with the help of some kind of differential equations, and there is involved a fluctuation which is some kind of white noise. There was a conference on biocomputation in Italy organized by a friend of mine, Ricciardi. He asked me to give a lecture on white noise. Biologists understood that many nonlinear fluctuations can be expressed in terms of white noise.

I: Do biologists understand the mathematics?

H: Honestly speaking, I don't describe everything systematically in advanced mathematics and I don't include everything in the lecture. There are many applications in biology which are interesting. I think a systematic approach will achieve something good. I was once heavily involved in the work on polymers and I even wrote a paper on it with Okada and Kiho. But I don't have any other mathematical results.

I: What about other applications?

H: Let me summarize the applications within mathematics — not in probability, but outside of probability. One is in fractional functional analysis. Fréchet, Lévy and others discussed nonlinear functions, $\mathcal{L}^2$ functions essentially, not random, however, from the viewpoint of infinite-dimensional calculus. We can see very intimate connections between white noise analysis and classical functional analysis. Second is in harmonic analysis. There is a duality between $\dot{B}$ and $\dot{P}$ which is Poisson noise. I have recommended people to discuss the infinite-dimensional rotation group O_∞, symmetric group $S(\infty)$ and their subgroups to see their roles in infinite-dimensional analysis.

I: What are the future directions of white noise analysis?

H: I would answer in the following way. Many people are more interested in Gaussian noise, but Poisson noise is also interesting. In the linear additive process, the noise can be decomposed into two parts, Gaussian and compound Poisson. They can be discussed separately. One may think that Poisson noise can be similarly treated to Gaussian, but I claim that dissimilarity is more important. There should also be a duality. To the Gaussian case, we can associate the infinite-dimensional rotation group. To the Poisson case, my colleague Si Si has associated the infinite symmetric group. The future direction is to discuss the duality between the Poisson and Gaussian cases in terms of harmonic analysis arising from groups.

Another direction is to work on the foundations of white noise analysis. There is mathematical beauty to be found regarding invariance, optimality, symmetry, duality and others, which should be investigated. There are also connections with other fields — in quantum dynamics, quantum information theory (quantum probability), molecular biology. For the last field, we are still at the stage of case-by-case study. Random fields should also be investigated, hopefully, in line with white noise theory. An application is to the Tomonaga–Schwinger equations — our group at Nagoya has kept up the interest. Once I had a conversation with David Mumford, the very famous algebraic geometer, and found that he now has an interest in probability. Random fields appear on his homepage. We should revisit the ideas of Lévy and Itô. I have organized Lévy seminars in Nagoya; last year's was our fifth Lévy seminar, and we proposed a new program for his ideas.

Roger

Exceptional Lie Group Theorist

Roger Howe is well-known for his path-breaking work in the theory of Lie groups and representations and for his impact on mathematical education and pedagogy through his teaching, writings and active involvement in educational reforms. His research is also directed toward the applications of symmetry to harmonic analysis, group representations, automorphic forms and invariant theory.

He has a bachelor's degree from Harvard University and a doctorate from the University of California at Berkeley. He taught briefly at the State University of New York at Stony Brook and, since 1974, he has been at Yale University where he has served as Director of Graduate Studies in the Department of Mathematics and as departmental chair. He has held positions at the Institute of Advanced Study in Princeton, University of Bonn, École Normale des Jeunes Filles in Paris, Oxford University and Rutgers University, Institute for Advanced Studies at Hebrew University in Jerusalem, University of Sydney, University of New South Wales, University of Metz, University of Paris VII, University of Basel, Kyoto University, National University of Singapore, Hong Kong University of Science and Technology. He has been invited to lecture in many countries throughout the world. He is a member of the National Academy of Science, USA, the American

Howe

Academy of Arts and Sciences and the Connecticut Academy of Science and Engineering. He received the American Mathematical Association's Lester R. Ford Award for exposition. He is an exceptional research mathematician who also serves selflessly and tirelessly on national and international boards and committees for the advancement of mathematics and the improvement of mathematical teaching and education. Among others, he has been involved in the CBMS project on the mathematical education of teachers, AMS Review Group for revision of the NCTM Standards, NRC Mathematics Learning Study and AMS Committee on Education. He was on the Board of Directors of the Connecticut Academy for Education in Mathematics, Science and Technology and the Mathematical Sciences Education Board at the NRC. Recently, he received the 2006 American Mathematical Society Award for Distinguished Public Service.

In recognition of his distinguished scholarship and exceptional teaching, he became the first incumbent of the Frederick Phineas Rose Professorship in Mathematics, and he was recently appointed the William R. Kenan Jr. Professor of Mathematics at Yale. His influence on his students is well recognized. In particular, his influence is strongly felt in Singapore in his role as Chair of the Scientific Advisory Board since the establishment of IMS in 2000. He has also bequeathed part of his mathematical legacy to the Department of Mathematics in NUS in the form of a strong research group centered round a number of his returned PhD students. In his honor and in appreciation of his numerous contributions, an international conference was organized at NUS from 6 to 11 January 2006 on the occasion of his 60th birthday. When he was here as the guest of honor of this conference and also for the annual visit of the advisory board, he was interviewed on 7 January 2006 at a café near Swissôtel The Stamford. The cacophony of the surroundings and the cold from which he was then recovering did not dampen the passionate spirit with which he talked about mathematics. The following is an edited and enhanced version of the transcript of the interview.

Imprints: You did your BA in Harvard. What attracted you to Berkeley for your PhD?

Roger Howe: The main factor was that it was in California. I had spent my high school years in California and I still consider myself a Californian. Although I was in the east for four years, I really wanted to go back. Berkeley was the best-known place in California while I was there. In some sense, I was naïve. I didn't even think about Princeton. I didn't know that Princeton was the place you might want to go to. In some sense I should have stayed in Harvard because I had won the Putnam Competition and that included a fellowship to study at Harvard. I think that some of the faculty there were somewhat shocked that I decided to leave, but I was very intent on getting back to California at that time. A more substantial reason for going to Berkeley was that it had both a very large and very strong faculty so that you can study almost anything.

I: Was anybody there whom you particularly liked to work with?

H: I had already gotten interested in representation theory, in which I have done most of my work. George Mackey who was at Harvard (I did a reading course with him in my senior year) had a student there [at Berkeley], Calvin Moore. He ended up being my advisor.

I: You mentioned the Putnam Competition. When you went to Berkeley, did you have a scholarship or something?

H: Yes, they actually offered me a pretty nice fellowship. There were some special fellowships from the government intended to recruit students into scientific areas and I got one of those fellowships. It was called an NDEA (National Defense Education Act) Fellowship. The NDEA was motivated by a desire to keep ahead of the Soviet Union in science. It supported many graduate students who went on to productive careers in mathematics and science.

I: Why did you choose to pursue research in pure mathematics?

H: Well, actually I hesitated a lot. It bothered me that there seems not to be more emphasis on connections. To me the applications of mathematics are a very appealing part of the subject and it is very important to me that mathematics helps you understand the world, but eventually I ended up going into pure mathematics and I have been very happy working there.

I: No regrets?

H: Not really, no. Well, yes, some; I wish I had done some more applied kind of things but it hasn't worked out.

I: Do you consider yourself to be an analyst first and then an algebraist?

H: Well, this is going to sound kind of funny but I actually consider myself to be a geometer. I kind of think geometrically. I've never published any work specifically on geometry but I love the subject very much. Euclidean geometry was one of my favorite subjects in high school and I've continued to think about it. I'm working with a colleague on a textbook for geometry. I think in terms of pictures although my work doesn't seem to have much to do with geometry.

I: Not many mathematicians think geometrically, they think more symbolically.

H: I think topologists often think in terms of pictures. Algebraists and analysts probably have other means of figuring things out. For some topologists at least, pictures are very important.

I: Your work is connected to some kind of topology, isn't it?

H: Lie theory actually — this is an attractive aspect of it to me — it connects with all branches of mathematics. It is geometrical but it is also algebraic, it is also analytic. Some of the main examples of geometric figures of manifolds and constructions come from Lie theory. That's a very attractive part of the field to me. You can connect things.

I: Are you a theory builder or a problem solver?

H: I like problems. I often will work on problems but I have never published a paper that solved a specific problem. And I very much like making connections, sort of coordinating things and connecting things together. On the whole, I guess that I'm a theory builder.

I: To be a theory builder, one has to know a fair number of fields to see the connections.

H: It helps to know different subjects. I have to say that I'm quite surprised how useful many things I learned for no particular reason have turned out to be… you learn things that don't seem to have connections with one another, and later on, you do find that there is some way to relate them together. That's very satisfying.

I: Algebra, or for that matter analysis, has its origins in rather "concrete" problems but modern research in these areas seems to be getting more and more abstruse and esoteric. Do you think that this is a desirable trend?

H: I think there's a constant kind of dynamical dialectic between what seems to be very abstract and the more concrete things. A very dramatic example in recent years is Mandelbrot's exploitation of fractal geometry. The basic work that Mandelbrot has exploited or publicized was actually invented by mathematicians in the late 19th century and early 20th century, and they were for a long time considered to be extremely abstruse constructions that could never have anything to do with reality. These were prime examples of things that only mathematicians would ever think of. And then Mandelbrot came along and said, "Actually, clouds, coastlines and trees and many, many different shapes in Nature share some of the qualities of these structures, and we can learn about Nature by thinking about these seemingly very weird structures." There's always a pull between the abstract and the general and things that seem to be far from reality, on one hand, and very concrete things on the other. It's also the case that physics is very weird and physics had to go far beyond our basic intuition in order to uncover a more fundamental truth in Nature. I think mathematics is like that, and of course, mathematics is a major tool in physics.

I: Did Mandelbrot discover those things independently or did he already know about them?

H: He was aware of the earlier constructions. He was the one who was able to see that there are things in Nature that are like fractals.

I: It seems that the success of algebra is its ability to reduce problems to symbolic manipulation but that the ideas of analysis are more intuitive and their formulation often precede their justification. Do you agree with this view?

H: Of course, there are some famous examples of that, like the Dirichlet Principle which was used in the 19th century for a long time before it was justified. This may be an example of mathematical riches in a direction that we don't very well understand but when we come to understand them we can reduce them to more basic things. I also want to say that in Lie theory there is a very interesting dynamical interaction between the algebraic and the analytic. It turns out, for example, that a fairly important aspect of representations of Lie groups is that there are some natural functions associated with them called matrix coefficients. An important fact is that for many groups, matrix coefficients die off — they go to zero at infinity, and this has implications for ergodic theory and counting rational points on various varieties. The proof of that is a very interesting interaction between the algebraic structure of the group and the analytic structure of some vector space. Again and again in Lie theory, you find these things, which you think of as different, interacting in an interesting way.

I: In general, geometric intuition seems to be very nebulous and often initially the ideas do not have rigorous justification.

H: It's hard to pin down, yes, but then you can spend very profitable, maybe very long, periods trying to figure out what are the reasons why this thing is true and you learn a lot during that process.

I: It's interesting what you said earlier on that you think in terms of pictures. That means that your intuition is basically a kind of geometric intuition.

H: I think it is.

I: It's quite rare, at least among algebraists.

H: It's hard for me to say. You can only know how you think. You can't know how other people think.

I: Could you briefly tell us some of the central problems in your area of research?

H: In representation theory in the strict sense, I guess the major fundamental problem still open is the classification of unitary representations. This has proved to be a very hard problem. Interestingly, the collection of all representations of a reasonable form has been known for 20 or more years but to figure which ones inside that set are unitary turns out to be a very hard problem. Then there are applications of representation theory to the theory of automorphic forms, and there, there are a huge number of problems which prove to be extremely challenging. A large part of it is what is called the "Langlands program" which has been challenging many mathematicians for several decades.

I: Has there been much progress in the Langlands program?

H: There has been very substantial progress but I'm not the best person to comment on it. In particular, Jim Arthur has established a rather general version of the trace formula and he has made applications of it. That's a good example where there is intuition and things are not proved, so there is a large web of conjectures. Only people who spend all their time thinking about them have a clear picture of what part is known and what part is conjecture. It's quite an amazing zoo.

I: How does it compare with the classification of the finite simple groups two or three decades ago? There were then a lot of things floating around.

H: That was a fairly well-defined problem. Of course, before the classification was achieved, people didn't know how far one would have to go. But the Langlands program is much more open-ended. I think that it includes problems that we, in fact, will never solve.

I: You mentioned the unitary representations. What is the significance of unitary representations?

H: Well, in physics, that is, quantum mechanics, representations arise because of symmetries of a physical system. And the representation should be unitary, because the states of the system come from vectors in a Hilbert space, and the inner product has a physical interpretation. In the theory of automorphic forms, again there is a natural inner product which was constructed in special cases before it was realized that representation theory was relevant. And of course, unitary representations have a nicer theory than more general ones, just as statisticians liked to use least squares approximation, because it is nice mathematically. However, not all representations useful in applications are unitary.

I: What are the prospects of settling this problem within the next 10 years?

H: Very good progress has been made. In fact, it is now understood that for any given group the problem of classifying the unitary representations comes down to a finite algorithm, but the question is: can it be made more specific to form some kind of global picture? Also, there are computational issues because for some of the exceptional Lie groups the computations that you have to do to carry out this algorithm may be very, very expensive. It's not a problem about which you know nothing. A lot is known. David Vogan and Dan Barbasch, in particular, have made a tremendous amount of progress, but it's not settled yet.

I: Has the computer been used?

H: Computers are being used. In fact, there is a group now working on setting up a website where you can go and input a given representation of a given group and it will compute for you whether that is unitary or not.

I: From your personal point of view, you would prefer something more conceptual?

H: More pictorial, yes. We need a more geometric picture of it.

I: What are some of the recent applications of your field to other areas?

H: The subject of matrix coefficients has applications to ergodic theory and counting of points on varieties (equidistribution of points in some larger space). That is one kind of applications. Of course, there is the ongoing application to the Langlands program, the theory of automorphic forms, where there is a constant interplay between representation theory and a broader spectrum of number theory. Recently, S. Alesker has used group representations to settle some outstanding conjectures in geometric integration theory.

I: What about to physics?

H: Well, this is what got people interested in representation theory at the start. Some representations of certain groups, you could say, are in some sense at the center of mathematics and at the center of the universe; in particular, a relatively simple kind of group called the Heisenberg group. Mathematicians call it the Heisenberg group because it is the group-theoretic embodiment of the Heisenberg canonical commutation relations in quantum mechanics. A tremendous amount of mathematics and physics relate to that group. Differential equations come in naturally and there are several absolutely fundamental mathematical structures which are connected with that group. There's the basic result of Hilbert's syzygy theorem in invariant theory or the linear algebra behind the Hodge decomposition of cohomology on Kaehler manifolds. So much is connected with this particular group that it's really amazing. Also, the quantum-mechanical hydrogen atom which is the basis of our understanding of chemistry is essentially a certain extremely distinguished group representation. So group representation theory in some sense is fundamental for our understanding of quantum mechanics. But it connects to many other things too. The odd thing is that these physical systems carry extremely special, very interesting representations of certain groups and the challenge is to find out what are the uses of more general representations. There have not been that many applications of general representation theory as we would like to have.

I: There are those super-Lie groups. Are they generalizations of the standard ones?

H: Yes, they are sort of combinations of several algebraic structures in one. Lie algebras are based on a product which is skew-symmetric — if you switch the order of two elements in a product, the product changes sign. There is another kind of algebra called the Jordan algebra which is commutative in the standard sense — you switch the order of two elements, the product doesn't change. Lie super-algebras are a combination of these two structures. They definitely have applications. They attracted interest and were classified when physicists became interested in so-called "supersymmetric" field theories. Actually, the Hilbert syzygy theorem involves a Lie super-algebra.

I: Is pure mathematics losing talented students to other more "glamorous" areas like mathematical finance or more applied areas like statistics and computer science?

H: This always happened to some degree. A talented person will have several areas to choose from and this has been going on a long time. Gauss had to choose between philology and mathematics. When he made some of his discoveries about the cyclotomic numbers and the construction of regular polygons, he decided that mathematics might be a better choice. Many people who could do mathematics can also do other things. Probably it's partly circumstance, what they get exposed to. I think it's also personality. For example, theoretical physicists and mathematicians tend to have very different personalities.

I: Talking about personality, there seems to be the observation that the personality of an algebraist is quite different from that of an analyst. Do you agree with that?

H: I would agree with that.

I: What do you think is the secret of your tremendous success in teaching mathematics at the university level?

H: Well, first, "tremendous success" are not words I would use. But I have worked hard to improve over the years, and it has been very rewarding to see students respond. Teaching is a complex art, and you can't sum up what you do in a few phrases. But the thing that I have worked on is to improve my communication with students. I spend a lot more time asking them questions, and less time just explaining. I sometimes say that I used to try to show students why math is easy, and now I try to get them to see why it is hard.

I: As the Chair of the Institute's Scientific Advisory Board during the past five years, what is your deepest impression?

H: I have been extremely impressed by Louis Chen, the energy and devotion which he put into this Institute. He has done a terrific job of soliciting proposals from the community in Singapore and trying to find ways in which IMS can help the mathematical community in Singapore. I think that without his leadership IMS would have been less successful.

I: I think you once mentioned the "critical mass" for active research...

H: This has been a problem and will be an on-going problem. Singapore needs expertise in mathematics and the IMS can help nurture that expertise and build it. If anybody can, Louis Chen will make that case to the Singapore government.

I: Of course, we will still need people like you to chart the direction.

H: It's really a pleasure for me to work with Louis and to help out the IMS. The whole Scientific Advisory Board has really worked extremely well together. I think we have done our best to be constructive and help make suggestions which might make proposals stronger. It's been a pleasure working here.

I: Do you foresee continuing working for IMS for the next five years?

H: Well, that's up to Louis and what he wants to do. It might be good to have fresh people in to get new ideas.

Wilfrid

Dancing with Randomness

Wilfrid Kendall followed in the scientific footsteps of a distinguished father (probabilist and statistician, David Kendall) and established himself as a well-known expert in probability theory who has made significant and wide-ranging contributions to random processes, stochastic geometry, stochastic calculus and perfect simulation. His interest in the use of computers in teaching and research has also led him to develop computer algebra software in statistics and probability.

He is on the editorial boards of numerous leading mathematics and scientific journals, among them *Annals in Probability, Statistics and Computing,* and the *London Mathematical Society Journal of Computation and Mathematics*. He has been invited to give lectures at major scientific meetings and conferences, and has served on the committees of international scientific organizations. He is a professor in the Department of Statistics of the University of Warwick.

He was the Chair of the Organizing Committee of the Institute's program on *Markov Chain Monte Carlo* (*MCMC*) held in March 2004. He was interviewed on 17 March 2004. The following

Kendall

are edited and vetted excerpts of the interview, in which he talks about the early formation of his career interest, the role of randomness in computer simulation, the close connection between probability and statistics and his views about the place of computers in statistics and intellectual thought.

Imprints: Where did you do your PhD and what was it on?

Wilfrid Kendall: I got my PhD, or DPhil as it is called there, at Oxford. I was an undergraduate at Queens' and then a graduate student at Linacre College. My thesis ended up with the title *Three Problems in Probability Theory*, which was very naughty of me because I had been told that the PhD title should be informative about what the PhD is about. But I was so anxious to get it submitted that I forgot all about the title until the last minute. There were, of course, three problems in the thesis. One was to do with early work on the knotting of Brownian motion, one concerned contours in random fields and one related to work I had done with my father on the statistics of shape. They were probability or statistics topics and they all had some kind of geometry attached to them, which has continued a fairly common theme in all the work I have done. My supervisor was John Kingman. He, in fact, was almost supervised by my father. Well, he was supervised by my father, but he never got around to submitting his PhD — never needed to! So my father is also my academic grandfather except for that small technicality.

I: Your father is a famous probabilist and statistician. How much were you influenced by your father?

K: It's a very interesting question to me. In one sense, enormously — the fact that he was a working mathematician, that research was clearly exciting and interesting for him. That had a great influence on me. On the other hand, he was very wise, and he knew then what I know now; that if you are following in your father's footsteps, then it can be a difficult path sometimes. And so he never pressed me at all. Occasionally he would tell me a little bit about mathematics but I never felt any compulsion from him to do mathematics or statistics. It was all a choice of my own free will. In fact, my free will was really well informed! At the time when I came to choose my subject for a PhD, my tutor at Oxford, whom I admire very much, warned me that it could be difficult to follow in one's father's footsteps. He gave me very sound, very helpful advice, and he said afterwards that I had listened to him very politely and then I went away and did just what I was going to do before. But I really did take what he said very seriously. However I found I hugely enjoyed doing not just mathematics but also probability and statistics. So I was led that way. I was doing it because it was interesting and engrossing. I didn't want to do it because it was something my father had done. I would be quite strong on that point to anyone in the same position as myself. You really must be sure that you are doing what you want to do because inevitably there are going to be times when it is difficult, and then you'll need to know that you made your own choice for yourself.

I: What is the difference between applied probability and statistics?

K: That's a tricky question! It's like asking what's the difference between strawberry and cream. They are very close, and it's really nice to have both of them together. In statistics, the questions are different: you are saying that there are things you want to know about, so you estimate and you test your hypotheses and so on. In probability, you are saying, “The system is behaving randomly and I want to know how it’s going to behave.” It’s a different kind of question. It’s going the other way, if you like. There is not a clear cut line of division.

I: What are some of the hottest topics or developments in applied probability?

K: I asked people at lunch about this question and we all agreed that this is a very hard question! Certainly I can give a personal answer: what is hot for me is all the things that I like doing at the moment. The whole interface between probability and computing is very interesting. What we are doing now (in the Institute’s program) is only a small part of that. There’s a lot going on. Some of the work going on in random matrices is absolutely brilliant, and there’s some lovely stuff to do with percolation theory. It is a difficult question to answer, particularly about applied probability. Some parts of science, and even of mathematics, are like a huge factory. You just have one or two products, and everybody involved is somehow working on the same products. It may take a long time before they eventually produce something really big. There are other parts of mathematics, and probability is one of these parts, which are extremely creative and vigorous, but there is no great master plan to which everybody tries to contribute a little. Instead, it’s a very rich and fertile field and there are lots of different problems coming up all the time with a lot of premium on being original and trying to find your own way to do things. Temperamentally, I find that much more exciting. But it’s difficult to say what the hottest development is in probability. You can say what you like to do right now but it’s probably unwise and counterproductive to try to have much influence on what everybody else chooses to do.

I: Or shall we say, what are the central problems in applied probability?

K: Well, I think there are central problems that people are looking at and getting intrigued by. I’m not sure if you should talk about probability problems. They typically are problems to do with mathematical science generally. For example, at the moment some of my friends are extremely interested in random matrix theory because they think it might have something to do with the Riemann Hypothesis. Sometimes people think there is something there, and sometimes people think it’s a mad dream. But it is interesting in its own right. There are other questions which have really been there a long time in statistical mechanics — whether there is some universal structure hiding behind things like percolation. There are people who have done some exciting work related to that. There are certainly big questions that people would like to think about. But I think that it’s true generally that’s what makes probability an attractive and vigorous subject, why a lot of people are attracted to it; there are lots of things to do and they are all very interesting. No one can quite tell what will be the next new development.

I: Could you tell us how the term “Markov chain Monte Carlo” came about?

K: Monte Carlo refers to the process where you want to calculate something and it may be too difficult to do either by hand or by using a computer to find the integral directly, and you try to do it instead by doing a random experiment, which involves the probability of interest. It actually goes back a long way — the famous Buffon needle problem. You drop a needle onto a lattice of lines. Find the probability of overlapping the lattice. (Clue: it is related to pi.) But Monte Carlo itself is a term coined probably during the war because of computational demands in the development of the atomic bomb. Why Monte Carlo? Well, because the method had to do with the roulette wheel and randomness. Markov chain Monte Carlo is a particular way of doing Monte Carlo. If you like, you could read it as "Monte Carlo with Markov chains". So when you are doing these random experiments, the question is how are you going to do the randomness? For example, like tossing a dice, tossing a coin, or running a roulette wheel. You may do it indirectly, you may say let's build a stochastic system, a Markov chain, and let's design it so that it has an equilibrium distribution which is what we are interested in. Then you run it for a long time and you observe the outcome and that gives you a way of handling the calculations. The adjective "Markov chain" describes a way of doing a Monte Carlo.

This idea goes back a long way, one of the first ideas people were using. There are many complicated problems for which the quickest approach is to relate them to the probability of long-run behavior of Markov chains. There was a very famous paper by Metropolis and others which goes back to the 1950s, but almost certainly they were doing a lot before that. The physicists who have a lot of money to buy big computers have always been interested in computing and developed it. Relatively recently, statisticians started to persuade people to buy them computers too. And the computers got flown in and sit on the statisticians' desks. At that stage, a large number of statisticians started to get involved using computers. Once they have the computing power, then it started to become a more feasible way to solve problems. It is pretty effective and has a tremendous influence upon the way people are doing statistics now.

I: How much of the new developments in probability and statistics have been dictated or influenced by the advances in computer technology?

K: I think, a huge amount. Here is a very simple example: the sort of questions that I used to mark for undergraduate examination papers when I started lecturing have largely gone out of fashion because they had to do with hand calculations but now you simply use a statistical package. I think that had a very big influence on the sort of things one does because some things have become very easy. One no longer thinks about them. But then, that means you can pose much harder questions. Markov chain Monte Carlo is another kind of example; computations that would have been inconceivable without accurate computing power. And then there are other applications, not really applications, but problems stimulated by the presence of computers and computation. You get interested in different kinds of questions. Back at Warwick I have a number of people I like to spend time to talk to — many of them are in the Statistics Department, many in the Mathematics Department, and also a significant number in computer science because probability is now important if you want to understand how to analyze the behavior of computer algorithms.

I: What about the theoretical aspects? The computer is good for computations, but will it have any influence on the theoretical development in probability?

K: As soon as you know how to do something, that there is a possibility of an answer, then your theory changes because your theory is about how you do things and you have just acquired a whole new way of doing things. That means you need a whole new theory. You can trace that all through statistics. What people are interested in theoretically is very often driven by the things that they can already do on the computer, which suggests theoretical questions. And then people on the theoretical side are motivated to do new things.

I: There seems to be a prevalent faith in some kind of order underlying every random, if not chaotic, behavior. Do you see this as a new paradigm in science or even in mathematics?

K: It's a very old paradigm. For example, in the book of Genesis, God builds order out of chaos. I think the idea of order coming out of chaos is not particularly new! Indeed, the mathematical notion of chaos can be viewed as saying there is a randomness in the choice of initial conditions right at the beginning, but you only see it bit by bit as the system evolves. I don't think there is any real conflict between randomness and systems with a great deal of order. Adrian Smith once said that probability is about what you don't know. You make probabilistic statements about the things you don't know are happening. It is perfectly compatible with ordered complex systems. Some things you don't know about, maybe you'll find out bit by bit as the system evolves. You can even use probability to do it. In fact, we had a conference in Durham in the summer which was to do with Markov chains in the overlap in between many different areas. And one of the things that was very interesting to see is that the group of people using Markov chain Monte Carlo in statistics were often working towards the same end as people who study deterministic systems with no randomness whatsoever and who are finding that the theory of Markov chains is a good way to describe how the initial conditions propagate through the system.

I: Does probabilistic modeling require the design of a "perfect" random number generator or some similar "random process generator"? Is that achievable in practice?

K: The answer to the first part is "no", and the answer to the second is "probably not". Practically, what you need is something which generates random numbers which are good enough. You don't want a number generator that produces a periodic sequence 0, 1, 0, 1, ... That's not good enough. How good is "good" enough? It's good enough if it does what you want it to do. If it has done its job, then it's good enough. A lot of work goes into the design of an arithmetical random number generator. From time to time, it gets replaced by one that is thought to be better and sometimes one can indicate rigorously how good the properties of these random number generators are. Indeed we have just had an example in the workshop: someone was talking about the case where you can show, a bad choice of random number generator leads to errors in certain complicated Markov chain Monte Carlo calculations. So you have to be careful. There is no replacement for the computer in your head. You've got to think about these things.

Suppose you want to produce a perfect random generator. Maybe you go to quantum theory, but there are all sorts of ways that things can go wrong. For example, suppose you built it wrong. That's embarrassingly easy to do if it's of complicated design. I recall a friend of mine who tried to build random number generators using thermal noise. He said that it was going to be perfect. He set up the stuff which electronically converted the thermal noise into noughts and ones, and it had a subtle correlation in it. He showed it to me and we agreed, "It's wrong. There is not enough random deviation." Eventually he traced the problem to some subtle kind of electrical feedback.

This morning, somebody was talking about the design of generators of random bits based on a Geiger counter but they failed to take into account the fact that the Geiger counter worked better at higher voltages and there was a 24-hour period fluctuation in the voltage supply to the Geiger counter. So in a technical sense it wasn't doing the job it set out to do, producing more random bits at some times than at other times. You have to realize that in the black box you are using to produce a string of random numbers, there's probably going to be some factor there which you can't quite control and which you might have left out. When you take that perspective, then it doesn't seem so crazy, on the other hand, to use what we call a pseudo-random number generator using an arithmetical sequence because at least, you understand the properties of that. One of the criteria in the practice of random number generation is that you should prefer a random number generator whose defects you know to one whose defects you don't know. There is a nice quote about this. John von Neumann said back in the '50s, "Anyone attempting to generate random numbers by deterministic means is, of course, living in a state of sin." You have to do it, you are using a random number generator, at the back of your mind there may be something wrong with the generator, or maybe it's something wrong with the way you code the thing. One of us was just estimating coding error probabilities this morning. He reckons that the programs he writes have a 40% chance of being wrong in a first working draft. For my programs the chance is probably higher. Once you take that into account, you start looking for the bugs you *know* must be there!

I: Are there any limits to the levels of computer simulation? Do you think that computer simulation can shed some light on some of the mysteries of life such as the way the brain functions or even the origin of life itself?

K: The answer to the first question is: "Yes, there are limits". The answer to the second question is clearly yes and clearly no. The first question is interesting. My friends in computer science tell me about some very interesting theorems which show that there are practical limits to what we can do with computer simulation and which are related to algorithmic limitations to do with the phrase "NP-complete". You are looking at a world of problems of scale. In other words, when you double the size of the problem, does the work you do double or quadruple or worse... or much, much worse... and hence you can derive notions of hierarchies of difficulty of algorithmic problems. You can get the same sort of hierarchy for problems to do with computer simulation. So there appear to be logical limits as to how much can be done with computer simulation.

Now to your second question. Science certainly can shed much light on amazing things. Everyday, for example, I read about new progress in understanding and control of diabetes. On the other hand, you just have to look into the eyes of a newborn baby to realize that there is something about which science remains silent. If, by the mysteries of life you mean something like that, the answer is: No.

I: Could you tell us something about the role and position of computers in mathematics education at your university?

K: Our department was one of the early UK statistics departments to use the computer in a big way in teaching statistics, so we were early starters. At Warwick, we have a center which tries to encourage innovation in the use of computing and it has taken on a very practical strategy. It recognizes that there are people using computers in all sorts of different ways across the university. It produces a newsletter which reports on these ways. It encourages people to experiment a bit and to report what is useful. Now, for example, whenever I give a talk or a course, I make sure that my lectures have notes on the web which are highly hypertexted so that they have all sorts of links in them. Increasingly, people say they like them and find these helpful. But I think that, while innovation and experimentation are good things, it's important always to bear in mind that actually education is ultimately about what is going on in people's heads.

I: Is it compulsory for Warwick statistics students to do some computer programming?

K: As a matter of fact, it is. Our students are all exposed to a course using the computing package Mathematica. But the point I'm making is that in the end what matters is when people walk out of the classroom or computer room, have they changed their way of thinking? Have they actually learned anything? You don't need a computer to make a difference to that and sometimes the best thing we can do to help people learn is to put the computer in a quiet corner of the room and switch it off. What matters is what's going on in people's heads.

I find the computer a great aid in making illustrative material available to students when I talk on some topics. It makes a tremendous difference if they can actually learn how to do things and see them afterwards. However it's important not to get lost in all those technology. We teach our students to use computer packages rather than programming as in such flexible packages you can learn how to program. We don't teach them programming as a primary activity. Typically, when they come out into the world, what they need to know is how to use the computer as a tool. That is clearly the way things are progressing. Programming is done by some people but what is most important is for people to know how to develop the qualities of systematic thought and care that are required for programming.

I: Do you have any connections with the Warwick Mathematics Department?

K: Yes, I have a lot of friends there. In particular, the Warwick probabilists are almost equally divided between mathematics and statistics. Probability is at the boundary and it is a good and interesting place to be in.

Lawrence

Economist for All Seasons

Lawrence Klein is a pioneer in the creation of computer models for econometrics and economic forecasting using mathematical techniques. From the formative years of his education at University of California at Berkeley and at MIT during the early war years, he moved to the Cowles Commission for Economic Research (then at University of Chicago) where he formulated a model of the United States economy and predicted an economic upturn after the war. He is well-known for the enhanced economic model called the Klein–Goldberger model and for the famous "Wharton Econometric Forecasting Model". He has built economic models of the United Kingdom, Canada, Japan and other developed and developing countries. He has served as a consultant to the governments of many countries; in particular, to China as it opened up to the West. The impact of his work on modern economics and his influence on present day economists are well recognized. He was awarded the Nobel Prize in economic sciences in 1980.

He was President of the Econometric Society and of the American Economic Association in the late '60s, and founded Wharton Econometric Forecasting Associates (now Global Insight) in the

Klein

’60s. In the ’70s, he started “Project LINK” to connect the models of some international countries in one of the first attempts to produce a “world economy” model. Besides being a practitioner, he is a scholar who has written extensively on econometric and economic models.

He joined the Department of Economics at University of Pennsylvania in 1958 and became the Benjamin Franklin Professor of Economics and Finance at its Wharton School of Business in 1968. He is now an Emeritus Professor at Pennsylvania and continues to be active in research and consultation.

He was a key speaker at the program organized jointly by IMS and the School of Economics and Social Sciences of SMU in April and May 2004 on econometric forecasting and high-frequency data analysis. He was interviewed by Yu Kiang Leong and Kim Song Tan, the latter of SMU, on 6 May 2004. In the following edited and vetted excerpts of the interview, the 84-year-old distinguished economist talks about his early university education, the scientific challenges of economics and a life-long dedication to the application of economics for the welfare of humanity.

Yu Kiang Leong: You mentioned in your Nobel Prize autobiography that you were attracted to mathematics and economics when you were in college. What made you decide to do your PhD in economics rather than in mathematics?

Lawrence Klein: For one thing, I started thinking about a long-term career as I entered college. I thought first about being an economist, and I stayed with that. I thought of mathematics as a tool to gain better understanding of economics. Also, when I was very early in university, we used to go out in teams to different colleges in the area to participate in mathematics competitions. I decided that I wasn’t really going to be good enough a mathematician to win those competitions and that there were young people of my age who were better mathematicians. So I stayed with economics.

L: Was your early mathematical training sufficient for your later work in economics or did you pick up most of the required mathematical methods as you went along in your research?

K: As an undergraduate in university, I had about half my classes in mathematics and about half in economics. But when I went to graduate school, I went further in mathematics. It was a very good graduate school at MIT and the mathematics was very good. So I picked up more mathematics and this was a period when mathematics was just beginning to be used in economics on a bigger scale, while most of the early work followed what we might call classical methods. When John von Neumann and Oscar Morgenstern introduced the theory of games, one had to go to set theory and other kinds of mathematical reasoning. I made the shift. And a lot of work dealing with dynamic systems in economics requires stochastic studies of dynamics that involves stability properties of systems and differential equations of more complicated sorts; so that I had to, for some time, keep studying mathematics. But then I got more and more involved in the applications of mathematics to economics and in the applications of economics to real world problems. Gradually over the years, mathematics got more and more complicated and deeper for economics and professionals. So I didn’t keep up studying mathematics endlessly. I shifted more towards doing things with economics and the mathematical basics that I had already started.

L: Do you think that mathematical rigor is necessary for economic training?

K: Rigor is important. You could be wrong, thinking you are right by not being quite rigorous and finding that there are exceptions and things that you have missed. I think what one really wants is imagination above rigor. And then you go to your friends who are mathematicians and check to see if your imagination and intuition took you in the right direction.

Kim Song Tan: In that context, how do you find the current trend in economical study which a lot of people complain is becoming too mathematical and not relevant or practical enough?

K: I won't say it's too mathematical but it's often too abstract. I think that some of the theoretical work in mathematical economics has drifted away from the important problems.

T: Do you see that as a potential problem in the sense that those who are in the university teaching economics are so well trained in mathematics that they feel compelled to continue along this path and go further and further away from the real problems that they might see?

K: They really do, but, of course, we have an important obligation to teach and give students ideas about economics. We should keep in mind that we are doing economics and not pure mathematics.

L: How much were you influenced by the style and philosophy of Paul Samuelson in economics?

K: That was important for me. When I was an undergraduate, I independently had the interest in finding how mathematics could be used in economics as a tool. One day I went to the library at the University of California and was thumbing through issues of *Econometrica* when I found in early issues (just beyond the first decade of the journal) articles by Paul Samuelson. I was so impressed by them that when I had a chance to go study under him at MIT, I realized that he was at that time, and still is, the greatest American economist of that period.

T: There are also people who say that people like you, Samuelson, Solow in your generation, are really great thinkers, not just economists, whereas the current crop of economists are just economists, technicians. Do you buy that argument?

K: The people you mentioned and some others like them are interested in mathematics to deal with the problems of the world we live in and to quantitative economics as a tool, but there are many people today who get away from that concept.

L: It seems that economics is, in some sense, observational and empirical and yet one cannot conduct controlled "economic experiments" in the sense of experiments in the physical sciences. Does it then make sense to talk about "economic truths"? If it does, are there objective "laws" in economics in the sense of scientific laws in physics or chemistry or even biology?

K: Of course, physics, chemistry and biology are very different. The broad concepts of science include some sciences that are very respectable but have no controlled experiments: meteorology (it's not an experimental science), seismology (it's hardly an experimental science) and astronomy. Yet they go far with mathematics. Apart from meteorology and seismology, astronomy is *very* precise.

Now, in addition to controlled experiments, which are important, the defining thing is the ratio of noise to signal. In astronomy, the noise to signal ratio is very low. Meteorologists have gone very far, but if you judge meteorology by looking ahead as far as one month or more, the findings don't look impressive. But if you judge meteorology by the next minute, the next day, or two days, it looks impressive and it's getting better. Economists should follow some of the techniques that meteorologists use. They tie in to the computer much more intensively. They send balloons into the atmosphere, fly aeroplanes through the hurricane's eye and learn more. We don't get enough of that extreme information flow in economics. Seismology understands what happens during an earthquake but they don't understand how to control or predict it. Now to some extent people are trying to introduce controlled experiments in economics. I often thought about that issue, say, like going to an institution such as a prison, change the economic values and look at the outcomes. It's possible to have some experimentation in economics. Nobel Prizes were awarded to some economists who did experimental work and went further than collection of data. But, by and large, economics is not an experimental science, and we must try to do the best we can with that limitation. The lecture I have just given was an attempt to show how we might improve our ability to forecast the economy by small steps even though we can't experiment.

L: Who were the economists who were given Nobel Prizes for their experiments?

K: Vernon Smith of George Mason University. He shared the prize with Daniel Kahneman of Princeton.

[*The citation for Smith reads: "for having established laboratory experiments as a tool in empirical economic analysis, especially in the study of alternative market mechanisms." – Imprints*]

L: Economic phenomena appear to be governed by random decisions in human behavior in a way not unlike those encountered in the history of mankind. Yet economics, but not history, has been hugely successful in making economics into a science. Is this due to a gigantic leap of faith on the part of economists in the methodology of mathematics?

K: In the work that I do in economics and econometrics, when something very big happens, like the OPEC decision in 1973 to limit oil production, to limit oil exports, to raise the price of oil four-fold or eight-fold, it is almost an arbitrary decision, unexpected. Now I say we cannot predict that OPEC would do that, but once OPEC has done that we can predict the outcome. I think we did very well with that. A number of the predictions that I have been involved with were of that sort. During the closing days of the Second World War, I shifted from MIT and was asked to help build a model to predict whether the United States will revert back to the Great Depression as soon as demobilization and peace were achieved. We made such a prediction. It was against almost everyone else's view, and it turned out to be right: America *would not* go back to the times of the Great Depression. And there have been similar times — after the Korean War, the Vietnam War, and even now, the present war. So we say we can't do a really good job in predicting those events but when those events have occurred we can do a reasonably good job in judging the outcome.

L: Are there such things as economic laws? Do economic laws exist?

K: I wrote a paper once on *Some Laws of Economics*. One of the interesting laws I looked at from time to time is called "Engel's Law", which he (Engel) found by studying social groupings of people in Europe — the percentage of a family's income spent on food declines as income rises. It may be a fairly weak law, but it holds. At the 100th anniversary of Engel's Law, the econometrician Professor Houthakker wrote a paper surveying countries all over the world to see if Engel's Law held. And then there was a very important event. When it came to China, he said he couldn't get data from modern China (that is to say, the beginning of the communist regime in China) but he found some Chinese family budgets from around 1920, or so, like the ones Engel found, and he said, "Yes, Engel's Law held." I was fascinated by those remarks. When I went to China for the first time in 1979, I got hold of a paper by a Chinese American economist surveying consumers in Tianjin and he got almost the same coefficient that Houthakker had found from the '20s in China for Engel's Law. So that particular observation by Engel had great longevity. There are many others that I cited like that. They don't give you enough information to know as much as you want to know about the economy. They don't cover a big enough part of it. There are some laws like that which have held up through centuries or decades.

L: When was Engel's Law formulated?

K: In the 19th century, in 1857. The people in the sample were Belgians, and Engel (Ernst Engel) was German or Prussian.

L: To what extent do the non-quantifiable elements of politics and culture contribute to the economic modeling of a country?

K: They contribute a lot. My last example in the lecture that I just gave was the use of sample surveys, not completely non-quantifiable, but not very quantifiable, of people's attitudes after the attack on the World Trade Center. The way we use surveys for consumers: to determine if you were better off, worse off (on a 5-point scale), much better, much worse or about the same. Let's say there is limited quantifiability. We found these very important in giving us guidelines on what consumers were going to do after that big event. There are many such things like that, and I claim that it's important to study subjective attitudes in decision making, political structure, and legal structure of politics and culture. Yes, they contribute. We should be aware of them and we should take them into account to the extent possible. Sometimes that extent possible can be stretched because we learn new methods of finding out about political and cultural events.

L: You constructed economic models for several countries like United States, Japan, Canada, United Kingdom and others. Were these models used by the various governments in planning their national economic policies? Have you done any economic model for a developing country?

K: The answer to the first question about whether they were used: yes; many governments have used these models or models evolved from them. I don't think you should use the word "planning"; I think it's in formulating their economic policies, such as interest rate policy or tax policy. It is not planning but it is using the models for doing the government's work and definitely used in that respect.

Any economic models for developing countries? Yes, for many. Right now, I have been involved, for a number of years, in modeling of China and I'm working on models for Russia. Russia is a transition country moving from planned economy to market economy. I have worked on Mexican models a great deal. I've helped a lot with different African models and different Asian country models.

L: Do you think that there is still a gap in communication, if not in interaction, between the majority of economists and the majority of mathematicians?

K: There are definitely gaps. If mathematicians are broad-minded enough and the economists are patient and careful enough, we still communicate quite well. But at the extremes, there have been major debates in our National Academy of Sciences whether the social sciences (that includes not only economics but also political science, sociology, anthropology) — whether some of them — should be included in the National Academy of Sciences. They have been there for a long time, but the academy had to make a conscious effort starting in the '60s or '70s to open up class groupings for social scientists, and some mathematicians have been very harsh in complaining about that. There were fights in the academy with mathematical members over the election of some social scientists. There was a very big fight at the Institute for Advanced Study in Princeton, over the hiring of a sociologist, by some mathematicians at the Institute. So these things happen.

L: Do you think that a mature mathematician could learn enough economics to make a non-trivial contribution to economics or do you think that he or she should possess some innate "economic acumen or intuition" in order to do so?

K: I don't know if it should be innate but I think that if mathematicians want to comment on the role of economics in social and political life, or economics as a social scientific discipline, then they should learn something about the way the economy functions and the way economic decisions are made, and then there will be better communication.

L: I will wrap up my questions with one question that is quite philosophical. You have dedicated your life to creating and developing a whole generation of economic models. Other than the Nobel Prize, what is the greatest satisfaction that your life-long work has given you?

K: Well, to see the models used. For me, one of the great things was that when I started in the faculty in the University of Pennsylvania, we knew the models and the application of models by the business and the public and government communities, and in doing so, we raised enough financial support so that over the years, between 10 and 15 PhD students every year were being supported by us. We paid their university fees and living expenses, and now they have gone out into the world and many have been very successful. That gives me a lot of satisfaction. We were able to use our approach to apply economics using mathematical, statistical, numerical methods to support enough students so that they have successful careers.

T: In your view, do you think that economic theorizing has reached a fairly mature stage, or do you expect to see another revolution coming in the same way that the Keynesian revolution, rational expectations, real equilibrium studies changed the thinking of economists?

K: I think that there is plenty of open room for creative thinking of a major sort to come and I think that the Keynesian revolution was very important. I think Leontief's work put activity up a bit; that is very important; and Samuelson's work was partly Keynesian. There were many others like that. But I feel that some of the work being done now is not getting far. I think that economists accept "rational expectations" as though it is realistic and correct; it is a hypothesis and I don't think it has been validated. There are others. I don't see a big event or a big change in the way of thinking among the most modern branches of economics that have the same impact as the Keynesian revolution had.

I think the information technology revolution had a very big impact, certainly a very big impact on what I do. It's not an economic theory, but it enables us to judge economic theory and principles much better. That was the basis of the lecture that I just gave. How can we improve economic forecasting by drawing upon the computer, the flow of information, the dissemination of information and the dissemination of policy preferences? There may also be a breakthrough, eventually, using the new kinds of techniques and facilities in the same way that I claim that meteorologists have definitely added, one day, two days, sometimes just half an hour, to the validity of meteorological forecasts. It helps utilities, helps the airlines, helps state planning. The economist will use the same information facilities to develop more accurate judgments and predictions.

T: You were involved over the years in work on China. From our understanding of economics, do you think that China's becoming a super economic power on par with America is a certainty? Can we say that it is a certainty that the Chinese economy will be a super economic power equal to the US?

K: I think you can say that China's catching up to the United States by aiming at a moving target would be unusual. I don't see China overtaking on a per capita basis eventually. I won't say it's impossible but it's not my judgment. On the other hand, China's present projection, I regard as plausible. China's leaders say that since reform (since 1978) China more than quadrupled in GDP by 1980. The new target is to quadruple again between 2000 and 2020. In looking at that, I'd say there's an excellent chance of doubling by 2010. I don't say they won't double between 2010 and 2020, but they will have to work harder than they have. I've been combing records. No country has had 40 years of that size growth in terms of established statistics. One reason why I think the decade will show whether it is favorable for China's plan is the preparation for the Olympics in 2008 and the preparation for World Expo in 2010. I think that those are going to keep China very busy providing the infrastructure and facilities for those major events, and China wants to show the world what she can do in those events that will give China the opportunity.

Brian E.

Modeling and Harnessing Turbulence

Brian E. Launder made important contributions, both experimental and mathematical, to fluid mechanics and convective heat transfer and pioneered the use of mathematical models to study turbulent flows. He was at the forefront of the development of numerical methods for turbulence models. He has also applied his methods to industrial problems related to turbomachinery and was active in leading research on environmental issues. His influence in engineering is extended through his wide and deep collaboration with his numerous students and other researchers.

He had his university education at Imperial College and MIT. Except for a short teaching stint at University of California at Davis, he taught mainly at Imperial College and from 1980 onwards, at University of Manchester Institute of Science and Technology (UMIST), where he variously headed the department, the Thermo-Fluids Division and the Turbulence Mechanics Research Group. He was Chairman of UMIST's Environmental Strategy Group, Director of the Mason Center for Environmental Research and Regional Director of the Tyndall Center for Climatic Change Research.

Launder

His scientific contributions were recognized by leading professional and scholarly bodies, and he was elected Fellow of the following professional and scientific bodies: Institute of Mechanical Engineers (IMechE), American Society of Mechanical Engineers (ASME), Royal Aeronautics Society (RAeS), Royal Society and Royal Academy of Engineering. He was Editor-in-Chief of the *International Journal of Heat and Fluid Flow*, an assessor for leading French institutions and advisor to Stanford University's Center for Turbulence Research. Though recently retired from UMIST, he continues to play a leading role in the Turbulence Mechanics and CFD Research Group.

He took an active part in the Institute's six-month program (July–December 2004) on turbulence. When he came to the Institute for a second time, he was interviewed on 16 December 2004. The following is an edited account of an illuminating revelation of his thoughts about a life-long fascination with an awe-inspiring physical phenomenon that is mysterious and gradually beginning to be fathomed and understood, if not harnessed.

Imprints: Do you consider yourself to be an engineer or an applied mathematician? Do you carry out laboratory experiments?

Brian E. Launder: Yes, I am an engineer but one who always enjoyed mathematics. When I applied to university at the age of 17 or 18, it became a matter of choice whether I became a mathematician or an engineer. I decided that my mathematics was good but not sufficiently brilliant to be a stellar mathematician. I thought, and I think it was a correct choice, that I could contribute much more to engineering at an applied level than by following a mathematics course. On the second question: as an undergraduate, I had a final year project in boiling heat transfer and that was experimental. I was fascinated by that. I wanted to do that for my doctoral study. So I applied to MIT to do my doctoral work and was offered admission to do boiling, but the gas turbine laboratory at MIT offered me a research assistantship which meant that I didn't have to do teaching-assistants' duties to earn my living. I would be paid to do research but, of course, the gas turbine lab would be concerned with gas turbines. From that point on, I forgot about boiling heat transfer and I concerned myself with the types of flow that arise in gas turbines.

I: Would it be correct to say that subsequently you were more theoretical?

L: No, even at MIT my work was experimental. It was certainly related to turbulent flow. So it was there that my interest in turbulent flow arose. But later on, I did move to mathematical modeling of turbulence.

I: Your CV mentions a number of doctorates to your credit, could you tell us a bit more about your graduate training?

L: As I have just mentioned, I decided to go to the USA for my graduate work. I applied to half a dozen institutions there. I had no idea what was a good university and what was a bad university in those days. I got several offers — one from Princeton, I recall, and one from Yale, and then an offer from MIT came through. I was advised by the professor at Imperial College to take the MIT choice. I did my master's and doctorate at MIT and then I came back to join the staff of Imperial College as a lecturer.

I: It appears that you have more than one doctorate.

L: Ah, yes, let me go on to say that three of the other doctorates I have are what we call higher doctorates. They are awarded on the basis of substantial body of published research — the first of those from the University of London I think I obtained in about 1976 when I was working at the University of California. I have to say that I got it purely to increase my salary. I don't know if they were impressed but I did get a small increase in salary as a result. Of the other two, one was awarded by the University of Manchester and the other by UMIST. In accepting those there was no effect on my salary. Nor did I expect any. I took them to show my commitment to Manchester. I submitted different research for each one and they were submitted at, roughly, 10-year intervals. It took me about a decade to accumulate enough new research to submit for a possible award of a higher doctorate.

I: Why did you go to MIT and not stay at Imperial College for your graduate work?

L: Well, I was interested in heat transfer and did think of staying at Imperial College, but the head of the Heat Transfer Section, Professor Brian Spalding, came and gave us a talk when I was in my final undergraduate year. Basically, he said, "Well, don't do your PhD here. The College is being rebuilt, you will lose a year of effort if you try and do your doctorate in this building at the present time." He was very honest and it is characteristic of him. He always says what he believes to be the truth. That stimulated me to look elsewhere. Since I thought Imperial was the best in England, if I wanted to look elsewhere, I naturally looked to America.

I: How did you become interested in turbulence research? Were you interested in turbulence right from your graduate days?

L: As I indicated, I got shifted into gas-turbine problems — problems of aerodynamics. I was given a free choice for my PhD project and I looked through maybe 150 alternative topics that the laboratory offered. The one that I chose I thought was very interesting. As an undergraduate, I learned a little bit about the transition of laminar flow to turbulent flow. I had always assumed what all the textbooks said: that was a one-way process; that is, once you get into turbulent flow, you never go back to laminar flow. But one of the projects that was put forward at MIT reported that the Russians had done some experiments suggesting that you could go from turbulent flow back to laminar flow if you accelerated the flow sufficiently quickly. This seemed a fascinating question to me. So I decided to look at it in more detail. The Russian paper had been complicated because the flow was supersonic and effectively they were looking at the flow around a projectile where there was a Prandtl–Meyer expansion wave, and effectively, in passing through the expansion wave, the turbulent boundary layer got peeled off and the laminar boundary layer grew up beneath. I was asked to look whether in subsonic flow one could get such a phenomenon. That was what I did for my master's and my doctoral theses.

I: Did it ever occur to you to become an aerospace engineer?

L: It did occur to me. Indeed, I did make some exploratory enquiries, when I was getting near the completion of my doctorate, about the possible positions. I didn't pursue that for two reasons.

Firstly, I was a holder of a Fulbright grant and if you hold that grant, you have to leave the USA for two years after completing your doctorate. I did think of going to Canada, but mainly I wanted to get back to England because my grandfather at that time had terminal cancer and I wanted to see him before he died. So having been offered a position by Imperial College, that seemed the best possible choice.

I: The Clay Mathematics Institute in the United States is offering a million US dollars for the understanding of the Navier–Stokes equations in fluid mechanics. Do these equations apply to all forms of hydrodynamic motion?

L: If one has a fluid that one calls Newtonian, that is to say there is a linear relationship between stress and strain, the Navier–Stokes equations apply to a large part of such flows. Of course, if one is in special regimes like free-molecule flows, then they aren't applicable. What makes them especially difficult to solve, however, is the nonlinear convective transport term in those equations coupled with the viscous term. Those two together are very challenging. Also, if the convective term becomes more dominant, then steady solutions, that is to say, solutions independent of time, no longer become stable. You get a transition to a phenomenon that they call turbulence. Occasionally, over a limited range, you can have periodic solutions, but the more usual form is this chaotic motion that is called turbulence.

I: Are the solutions analytically obtainable in principle?

L: Not by formal mathematics. Nowadays, computers are large enough that for a limited range of Reynolds number, one can with the computer solve the Navier–Stokes equations numerically. Of course, there are some analytical solutions for laminar flow.

I: Is turbulence a matter of boundary conditions?

L: No, even if you have perfectly calm inlet conditions, if the Reynolds number is high enough (high Reynolds number means that you have only got very weak damping by viscous forces), in that event you will find that small perturbations will grow and you will rapidly end up with turbulent flow even if you have a laminar flow coming in.

I: It's not that it is asymptotically so?

L: No, it is certainly not asymptotic. If you have a high Reynolds number flow (the Reynolds number is a measure of how strong the convection terms are compared with the viscous terms) and you have a laminar flow entering the domain, the smallest perturbation will trigger a transition almost immediately to a turbulent agitated motion. It's not asymptotic, it occurs almost instantaneously.

I: Is it also due to an initial condition, like a high Reynolds number?

L: No, it's simply the ratio between how fast the flow is going, what is the size of the range you are looking at, and what is the viscosity of the fluid. One is looking at things which are not related to boundary conditions.

I: In the real world, the phenomenon of "turbulence" seems to be the rule rather than the exception, and the human mind seems to be able to survive, if not thrive, under conditions of "turbulence". Why is turbulence scientifically so intractable?

L: [*Laughs*] I'm not sure how far I can answer this question. To start with your beginning question, turbulence is the rule rather than the exception in the world we live in. Why is that? It's because the two most important fluids to us (air and water) are, on the scale of fluid viscosity very small in value. If we lived in a world of castor oil, we would find that most of the fluid motions that concerned us would be laminar rather than turbulent. Above a certain Reynolds number, the flow does become chaotic. Unless one is going to make a major numerical solution of unsteady 3-dimensional turbulent flow, one is forced to look at the averaged equations and, because of the nonlinearity of the equations, one finds extra terms (what are known as Reynolds stresses). So, one is forced to adopt modeling to determine those unknowns.

I: Is there a fair amount of modeling for turbulence?

L: It depends at what level one works. I mentioned that one can, over a limited range of Reynolds number, solve the turbulence equations. If one wishes to do what I call "light modeling" of the equations, one can adopt what we call a "large-eddy simulation". This is in some way like a direct numerical simulation except that one recognizes that turbulence has fluctuations on a scale smaller than the numerical grid one is using and one has to include a model to represent that subgrid scale fluctuations. At that level, a simple turbulence model is the usual choice. Most of the effort is in the numerical solution. The model is just a small part of the numerical scheme that ensures that turbulence is destroyed at the required rate. If one goes to the level of modeling that I work at, however, (this is called Reynolds-averaged modeling), then there is a huge amount of input into the turbulence model because all of the statistical fluctuations are contained in the model of turbulence, whereas at the level of large-eddy simulation, most of the transport associated with turbulent flow comes from the simulation itself and it is only a small amount associated with the model.

I: Are the models you mentioned verified by experiments?

L: Indeed, one verifies them either by experiment or increasingly nowadays, by referring to large eddy simulations or direct numerical simulations. There are models that inevitably don't cover all turbulent flows. They will have a range of applicability. Modelers try to make the range of applicability as wide as possible. A one-flow turbulent model is of no good to anyone.

I: Feynman once said that turbulence is the most important unsolved problem of classical physics, and Heisenberg was reputed to have said that he had only two questions to ask God: "Why relativity?" and "Why turbulence?" How much nearer are we to a clear understanding of turbulence?

L: I'm not sure I can respond to this question in a meaningful way. I think that "understanding" is such a personal state, it's almost like religion. For myself, I feel remarkably assured when I am in tune with direct numerical simulation. If we can numerically solve the Navier–Stokes equations, it's very nice to see that the flow that comes out is what we see when we do a careful experiment for identical conditions. Now, at the level of modeling that I do, I find understanding in looking at the relevant equations (the Reynolds stress transport equations) and gaining insight from the different roles taken by certain terms under particular force fields or strain fields which explain why turbulence behaves the way it does. For example, why is it that, when we have a rotating flow, due to the resultant Coriolis force one initially gets some augmented turbulent mixing on the high-pressure side of the flow and diminished turbulence on the opposite (low pressure) side? One can see this directly just by looking at the equations... qualitatively at any rate. One can also understand qualitatively why, on the high-pressure side there is a cut-off level beyond which further increased mixing does not occur whereas on the low-pressure side mixing is continuously reduced until the flow becomes quasi-laminar. Yes, I find great insight and understanding of turbulence by looking at those equations. Of course these insights are usually not precise enough to give quantitative answers. One needs to provide a model for the unknown terms, to complete the mathematical specification of the equations and then, by solving them, one can get, hopefully, accurate results.

I: So you think that understanding of turbulence has increased tremendously over the last 50 years?

L: Undoubtedly so. I don't say there are no unresolved problems but there has been tremendous progress. Of course, as in many fields in the mathematical sciences, with the coming of the computer and the development of techniques to exploit the ability of the computer, it is inevitable that one sees many changes.

I: Is there a master plan or program conducted by researchers around the world to solve the turbulence problem like what the physicists are doing for, say, a theory of everything in string theory?

L: Turbulent flow covers so many fields from blood flow motion in the human body to atmospheric turbulence. You've got such a large range of scale. People's own interests don't span that range. I said atmospheric turbulence; gosh, stellar turbulence also has a large group working in it. So some areas may have this "grand plan"; it may seem grand but it is relatively focused and limited compared to the range of turbulence. No, there is no "theory of everything" in turbulence. Nor will there ever be, I suspect.

I: Will an understanding of turbulence provide the answers to problems in meteorology?

L: Of course, at one level. We are dealing with air which is a Newtonian fluid and it gives rise to turbulent motion. It is, in principle, described by the Navier–Stokes equations. No matter at what level we attack the problem, whether it is direct numerical simulation, large eddy simulation or the sort of Reynolds average modeling I do, what is so different is the scale. One is looking at thousands of kilometers; yet the smallest motion of turbulence is still as small as they are in the experiments I get involved with — they are fractions of a millimeter. There is really no way one can adopt the same approaches. So while people working in atmospheric turbulence do make some use of the approaches that we adopt, they are almost like a special large eddy simulation. A special feature of the atmosphere is that its horizontal extent is very much greater than its vertical extent, for example.

I: Recently I read in the *Scientific American* that some experiments done by a group at the Delft University of Technology in the Netherlands detected some kind of small eddies or currents that are supposed to be "building blocks" of turbulence. What do you think of that discovery?

L: Well, I'm not sure what words they used. Workers in turbulence, as in other fields, are always wishing to promote what they are doing by making it sound very general. Terms like "building blocks" are frequently ones that come to mind as an attempt to make a very complicated subject understandable by people without any specialized knowledge in the field. At Delft, there are some very strong people in turbulence. One is my first ever PhD student, Professor Hanjalic. He has been doing some fine work there, particularly on Rayleigh–Benard convection, in which one takes a pair of horizontal plates at some fixed distance apart. If one then heats the lower plate natural convection is started — that's what we term Rayleigh–Benard convection. You get something like that in the atmosphere. I mentioned Hanjalic because you asked me about the applicability of our model to atmospheric phenomena. He is originally Yugoslav and lived in Sarajevo that has a long valley. It's a city that suffers desperately from pollution. He has done a CFD study of the Sarajevo valley to help local authorities decide whether they should put a chemical plant with potential emission in one position rather than another. Clearly, one wants the effluent gas from the plant to be carried far away. On the atmospheric micro-scale, one is using just the same methods that we were using in engineering.

I: The designs of airplanes and ships are presumably related to the study of turbulent flows in air and water. Yet, Nature has provided its own designs in the form of creatures that fly and live under the oceans. Has there been any attempt to look for possible answers in the designs of Nature?

L: That's an interesting question. The answer is "Yes, we have". Let me cite a couple of examples. In the late 1970s and early 1980s, we (the community, not just myself) empirically discovered, principally from research funded by NASA, that by putting small longitudinal ridges in the surface you can actually reduce the drag to levels below what you would get on a smooth plate. (This seemed remarkable and it has been established and understood now.) Then people said to themselves, "Okay, we have gone this far. Can we do better?" because they were getting drag reduction of the order of 10%. Somebody called Bieter Bechert, who was a professor in Berlin and is retired now, looked at the performance of sharks. The great white shark seems to have an ability to swim at speeds faster than what people feel it should be able to, given its size and power and so forth. Bechert's idea then was that the shark must have some drag-reducing feature on its skin. He then took some shark's skin and examined it in detail and discovered that the shark had something like riblets — these devices that had been discovered empirically. But, of course, what the shark had weren't shaped exactly like those that have been evolved empirically. So Bechert launched a major research effort mimicking the shark's skin. Alas, I have to say that Bechert came to retirement before those experiments reached any firm conclusion.

I: Could he do better than Nature?

L: Looking at the results that he published, I don't think he did better than Nature. Looking at it another way, one can say that Nature had the same idea. I said I would quote two examples. So, as a second example, an area where it is certainly the case that one is looking at Nature is in the development of what I call "micro-sized vehicles". There is a lot of interest, partly from the point of view of surveillance but not only from that point of view, in building flying machines that are really the size of just my fist. Obviously, they will not contain anything except sensors: a small camera, sound recording equipment, or whatever. If you go down to that sort of size, you can't have a fixed wing. It is just not feasible. So various designs of flapping wing devices are being explored. Naturally, one looks extensively at the various solutions that Nature has come up with over the millennia in order to get good ideas.

I: But then birds are not the same as inanimate objects. Birds are "dynamic" but aeroplanes are not.

L: Well, there is a lot of research in this field (it's not an area where I do research), in bringing in basically dynamic response capability into inanimate objects. Certainly, it is the intention that smart devices should be embedded in, say, the wings or some other part so that if things are not exactly ideal, this can be sensed perhaps as a perturbation in pressure and there can then be feedback control to change it. However, what man creates in his instruments will be inferior to what Nature, even in a bird or an insect, has evolved over millions of years. Nonetheless, there is movement in that very direction.

I: Do you work on any projects in industry?

L: Well, I have done a wide range of industrially-driven problems sponsored by industry. Let me just mention a few. One has provided research, both experimental and computational, for at least 10 years — it's on blade cooling. It may not be generally known that the jet engine becomes more efficient the higher the temperature one makes the air flow coming out of the combustion chamber. The problem is that when it meets the turbine blades, the blades will melt at that sort of temperatures they want to use. So, one puts cooling passages in the inside of the blades. Because turbine blades are very small it is pretty challenging to develop effective cooling systems inside them.

I: Is the cooling done by coolant or by air itself?

L: It is done by air that has been compressed in going through the compressors and is then taken off before it goes into the combustion chamber. The air is heated up by the sheer compression but it's heated even more in the combustion chamber. This high pressure cool air is, I'm guessing here, around 600°C; that's a high temperature, but it is much cooler than the temperature on the outside of the blades. So that's what protects the turbine blades. So that's one area.

I'm also looking at the trailing vortices that are being created behind an aircraft wing. These are created at the tip of the wing as the aircraft flies through the air, and as you know, it can be dangerous for following aircraft to get caught up in these vortices. We are looking at ways to cause the vortices to die out faster. Finally, I mention research that I'm doing on nuclear reactors where one is trying to work with the people designing the next generation of nuclear reactors to improve better ways of cooling.

I: Has your research resulted in any patents or immediate applications in technology and industry?

L: Although I've been talking about my industrial work, I actually operate at a fairly fundamental level. So it isn't that I discover something that can be patented. So there's no patenting of my modeling work but there is a lot of industrial take up of it. My colleagues and I have advanced mathematical modeling so that it is used in industry. The turbulence model (not just my own work but work of modelers around the world) has altered what has gone into the computer programs that industry use. Much of the industrial computing actually makes use of what are known as the industrial codes. There are now three, perhaps four, major CFD (computational fluid dynamics) vendors. Certainly the vendors have imported my group's models into their codes. That is sometimes a terrible struggle, I have to say. Some code vendors have got models I produced thirty years ago and they're still using them. They complain that they don't work and I've been producing new models over the last 30 years. It's hard to get them to throw away what's in their codes and put in the new models.

I: It's very surprising to me that, given that engineers are well-known for going for patents generally, you are approaching your work from a more intellectual point of view.

L: It's probably what I'm best at and certainly what I enjoy. You will find people in the field who are more financially driven. I mentioned at the beginning the professor at the time when I was an undergraduate. He was probably the first person to get a serious CFD industrial program (that is to say, commercial); he was a commercially concerned scientist. Another colleague from my early days now has one of the most successful current CFD companies. These people just have different interests and different skills. They are more into the numerical discretization of the equations. I'm more into the physics. Like most academics I suppose, turbulence modelers are much more interested in our subject and the associated ideas than turning ideas into money.

I: The work you did on blade cooling must surely be patentable and could have been sold to aircraft companies.

L: Yes, but what we produced then is a software for computing a particular arrangement. What one could devise is an entirely different type of cooling arrangement — that would, I believe, be patentable. But that isn't what we were directed at. We were simply showing industry that we can compute this type of flow because we do both experiments and the CFD work. That enables the industry to say, "We'll use your computer code with its modeling and use that to design for ourselves a more efficient cooling system."

I: Do you have any PhD students?

L: I don't have any at the moment, and the reason is that I'm 65 and until a few months ago, I was going to retire when I reached that age. But my university has offered me an extension of my contract for two years. However, it seems to me irresponsible to take on PhD students at an age when I won't be there to supervise them over the final 18 months of their doctoral research.

I: What about in the past?

L: Oh, I've had over 40. Yes, that's the biblical number to signify "quite a lot". I still have some postdocs working with me — three postdocs at the moment. I also interact with academic colleagues helping them in preparing research proposals and offering advice when it's sought. I'm also under contract to write a book for Cambridge University Press. So, if I can get rid of all the administrative work that clogs up my days (and evenings and weekends!) I hope to make some further contributions to the modeling of turbulent flows. In that respect, the opportunity to contribute to the turbulence program here in Singapore has been a real pleasure.

Fanghua

Revolution, Transitions, Partial Differential Equations

Fanghua Lin (林芳华) is world-renowned for his important contributions to classical analysis and its applications to nonlinear partial differential equations.

Lin graduated from Zhejiang University, China in 1981 and obtained his PhD from the University of Minnesota in 1985. He was an instructor at the Courant Institute of Mathematical Sciences, New York University from 1985 to 1988 before going to the University of Chicago as full professor in 1988. He returned to New York University in 1989 and was awarded the Silver Professorship in 2002 by the Courant Institute, where he continues to produce outstanding research in the Courant tradition of hard analysis applied to nonlinear partial differential equations.

His research output includes more than 160 research papers and three books of lecture notes. Even before completing his doctoral studies at Minnesota, he had already made a reputation for writing many research papers. Among his many contributions are his fundamental work on the Ginzburg–Landau equations with a small parameter and his deep results on harmonic maps and liquid crystals.

Lin

His honors and awards are numerous, notably the Alfred P. Sloan Research Fellowship, the Presidential Young Investigator Award, AMS Bôcher Prize and S. S. Chern Prize (China). He was elected to the American Academy of Arts and Sciences and has been invited as distinguished speaker at important scientific meetings and conferences in the US, China and Japan and in major universities throughout the world. Besides supervising a number of doctoral students and postdoctoral fellows, he serves on the editorial boards of leading mathematical journals, such as *Communications in Pure and Applied Mathematics, Analyse Non Linéaire, IHP, SIAM Journal of Mathematical Analysis* and *Journal of Differential Geometry.*

Having established himself as a world leader in his field, Lin has not forgotten the moral obligation he has always felt towards China, his country of origin. For the past two decades or so, he has been actively engaged in promoting international scientific contacts and opportunities for the mathematical community in China. At a different level, Lin has a close association with NUS's Department of Mathematics and Department of Physics. He was co-chair of the organizing committee of the IMS program on *Bose–Einstein Condensation and Quantized Vortices in Superfluidity and Superconductivity* held from 1 November to 31 December 2007. More recently, he also served as a co-chair of the organizing committee of the IMS program on *Mathematical Theory and Numerical Methods for Computational Materials Simulation and Design* (1 July–31 August 2009) and of the program's Summer School (17 July–19 June 2009). He was interviewed on 10 December 2007 during the abovementioned program in 2007. The following is an edited and enhanced version of the transcript of the interview in which he offered a glimpse of a student's life during the Cultural Revolution that swept through China during the 1970s and how he emerged from the throes of that period as one of the first batch of students to enter the reopened gates of Chinese universities and how he was sent as one of the pioneering group of students to do graduate studies in the US. In the interview, he exuded a passion for research and open-mindedness towards learning in mathematics and science.

Imprints: You went from Zhejiang University to University of Minnesota for your PhD. Why did you choose Minnesota and what was the topic of your PhD?

Fanghua Lin: The answer is very simple. I didn't choose Minnesota. My professor at Zhejiang University chose it for me. That has a lot to do with the history of the department. The Mathematics Department of Zhejiang University is historically very important in developing Chinese mathematics and has trained a lot of Chinese mathematicians. At the time I entered university, it was February 1978. The department had only a few professors. But, the professors made a decision to choose the top 10 students of the classes of '77 and '78 and send them to US and Europe for their PhD and to get them to return to Zhejiang University. This was a rather strategic plan but as in many other schools in China, the plan was not carried out. Why would such an idea come up at Zhejing University at that time? It has to do with the history of the department. The early professors of the department, Professor Chen Jiangong and Su Buching, both got their PhD from Tohoku University, Japan around 1930. At that time, you don't see so many PhDs in China. With only two PhDs, it was already such a great department. So the professors thought then, if we have 10 PhDs, we will, you know, be even better. Because of this decision, I was chosen among the 10 to be sent to US to study partial differential equations. Going to Minnesota was also decided by the professors. The professor visiting Minnesota then was Dong Guangchang who was doing PDE, and the chair of the department then was Professor Guo

Zuorui. It's very hard to say I really chose the topic of partial differential equations. When you get to the US, you basically follow whatever you like to study. It turned out that my thesis topic was not on partial differential equations. It was more on a geometric variational problem and my PhD advisor was Robert Hardt.

I: That was in the late '70s?

L: I went to Minnesota in 1981. I entered Zhejiang University in 1978. I had basically three years of undergraduate study, the last half-year learning English mostly.

I: Was it on a national scholarship?

L: I got a teaching assistant fellowship from University of Minnesota, so I had to spend half the time working and half the time studying. It was kind of challenging but interesting.

I: That was after the Cultural Revolution?

L: Yes, I was in the first class (the 1977 class) that entered university after the Cultural Revolution, but the class entered university in the spring of '78.

I: Did the events of the Cultural Revolution affect your studies before you entered university?

L: Certainly, I never really studied at all in elementary school or for that matter in high school. Actually, thinking back, I liked it because it was completely free. There was no homework, no exams. There was no serious exam before I took the university entrance exam. The elementary school then was like a political camp.

I: How did you study your mathematics in high school?

L: I entered school at the third grade basically because the first two years was the start of the Cultural Revolution. We didn't really learn anything except Chairman Mao's quotations. At the fifth grade, we were starting to study the solving of equations. A lot of my classmate experienced difficulties with the problems, but I found them particularly easy. I found it started to get interesting because there was something else, not just something mechanical. My elementary school teacher thought that I did have some natural talent. I spent a year or two in elementary middle school (sixth and seventh grades) studying a lot of mathematics and physics too, basically by myself. I happened to meet a very good teacher. He was our physics teacher and he gave me the books published before the Cultural Revolution and some special books in mathematics. I read most of them and found them not too difficult to learn from. But then I really didn't spend too much time studying in high school because it was still during the Cultural Revolution and you didn't see the end of it. You didn't see much future then and therefore you didn't do much work — eventually, you would become a farmer. But I enjoyed the free time however, never really followed any rules or studied anything systematically.

I: Maybe it's not that bad for creativity.

L: Yes, in some way. Because of that I was always interested in thinking about problems and trying to solve them by myself instead of reading them in the books. Of course, there are advantages and disadvantages.

I: Except for a short stint at the University of Chicago, you are essentially based at New York University. What is the attractive factor of NYU?

L: Many reasons. First, New York City is very unique. You feel at home there. Everybody feels at home there. It has so much to offer: music, art galleries, museums, movie theatres, restaurants. It's just fantastic. I'm a lazy person. I always want to get things very easily — live near to food, get to work easily and everything should be very accessible. The other thing is that the Courant Institute is one of the greatest institutes in the world. You feel very warm and you find a lot of colleagues, so educated, and from different cultural backgrounds. I feel the friendliness.

I: Did you feel any culture shock?

L: I didn't feel much of a culture shock. I'm always very open. When I was a graduate student at Minnesota, I interacted mostly with students from other countries, a couple from Hong Kong. Of course, the culture thing is a much deeper matter. As time passes, I find I'm still very Chinese — some things never change. However, you don't feel like a foreigner at the [Courant] Institute. People simply respect you if you do a good job. The faculty is very happy when you accomplish something. People congratulate you and so on. This is a particularly friendly place. You don't have to prove to your colleagues that you are good or so excellent, which sometimes happens in some other places. I like other places too, like Chicago. I love the University of Chicago very much. It's a very English society — gentlemen, treat you very nicely. It's great. The weather is very tough, however, particularly at the time I was there. I enjoyed it very much and that is why later my family went back there again. I've been at Berkeley for half a year as a postdoctoral, Princeton for one sabbatical year and the Institute for Advanced Study for half a year as postdoc.

I: Have you gone back to China since then?

L: Yes, many times. My first return to China was in 1989 just before the Tiananmen Square incident. That was a cornerstone. That was my first visit to China after eight years in the US. A lot of things were changing and it was very nice. After that I practically went back to China every year and spent a couple of months; in recent years, always two months or more. I go back in summer to give courses to graduate students and try to find postdocs and good students. At the beginning, none of us who became professors would return to China, we would settle in the US. It's hard to say whether it's good or bad. But I think, in general, I'm still very positive that we can still go back to serve the country in some way. China has changed drastically over the last 10 years in particular. If it was 20 years ago, I would have gone back to China already. When I graduated in 1985, it was a very different time.

I: Your research spans both pure and applied mathematics. Were you already interested in applications to physics right from the beginning?

L: Physics is always a subject that I liked ever since high school. In the college admission test, I got my best score in physics. But I never really got seriously interested in physics. I always see myself as a mathematician. I am always interested in mathematical issues which may or may not be related to natural phenomena or science. After many years at the Courant Institute, my philosophy and point of view have been changing. To me, as your age grows, you realize that when you are young, whenever people tell you some problem or you see a problem, you just jump in and try to solve. But you gradually realize that there are simply too many problems, infinitely many problems, and "a man should know his limitations". You cannot solve all of them. And therefore, you have to be very selective. As age grows, your view changes and you spend more time selecting the problems. To me the type of problems is very important. At any point of time, there is only a small set of really interesting problems. When you look at the publications 50 years ago, you say, "Oh, why are the papers on such bizarre subjects?" You can be sure 50 years from now, people looking at our publications will say the same thing. What this means is that the problems that are interesting now may not be interesting in the future. So one has to choose a problem that is not just mathematically interesting but also relevant to what is going on in science. Science is always developing; it's not just imagination and creativity, it's driven by practical needs.

I: How do you select your problems you want to work on? Do you get them from journals or do you talk to people?

L: You choose a problem depending on your training, background and interest and by reading and talking to people — for me a lot of time reading non-mathematical articles in *Nature*, *Science* and so on. You know what is relevant and that is very important. Of course for some mathematicians, you can simply ignore what is going on in the world and do whatever you like. You know, science develops so drastically and is so diverse that if you don't pay attentions to the whole picture you are going to miss out quite a lot.

I: Courant Institute is mostly inclined to applied mathematics, am I right?

L: Yes, but we have faculty in both pure and applied mathematics, probably stronger in applied. But we have a very strong group in pure mathematics too. In the University of Chicago, I was regarded more as an applied mathematician, but at the Courant Institute I'm probably more on the pure side of mathematics.

I: Would it be fair to say that partial differential equations used in modeling physical phenomena are often based on simplified and ideal assumptions. As such do these equations actually reflect reality?

L: First of all, I would like to answer in a philosophical mode. Absolute truth or reality doesn't exist or is probably not so important to us. Even if it exists, when we try to understand natural phenomena, it's through our perception. So when we talk about reality or truth, it's always an approximation. If we know the absolute truth or reality, we probably understand the problem so well and therefore the problem is not worth studying. We model by using partial differential equations and other mathematical methods. A model is a model. Therefore, you have to simplify and make certain fair assumptions. But between different models one can sometimes distinguish between good models and not so good models. So what is the distinction? First of all, we want simple models because we can understand simple things better. If the model is as complicated as the real problem, what is the use of the model? A good model should always capture the essence and characteristics of the issues you want to address. A minimum requirement of a good model is to be real enough. How much you want is a practical trade-off for your needs. Yes, partial differential equations always use simple models. The good thing is that most of the time when we understand these models, we also understand the general situation.

I: Would you consider modeling more of an art rather than a science?

L: It's a bit of both. You cannot forget the fundamental issues you want to capture or understand; this aspect is science. How you do it — you do it nicely and elegantly or you just do your very best — that is art, and it also depends on the technology available.

I: Is there any recipe for doing good modeling?

L: I'm not a specialist in modeling. I think it's like doing physics or mathematics. Unconsciously, people use some very basic principles.

I: Which type of problems is more tractable? Evolutionary problems (i.e. parabolic or hyperbolic type) with given initial values or elliptic boundary value problems?

L: It's very hard to make a distinction or comparison between different types of problems and say one is easier or more difficult than the other. The problem can be extremely difficult for simple questions. The problem can be relatively simple even though it addresses a complicated system of equations. Really, it is what you want to achieve. If you want to access a large space of phenomena, even for very complicated systems, it is easy to come by. If you want to understand very delicate, very detailed information of some specific issues, then you have to look into very detailed characteristics of the problem and it can become difficult to solve. I think the difficulty depends on the issues you ask or the final conclusion you want to draw from the problem. It is not the type of problem (stationary or evolutionary) posed that is easier or harder.

I: Are the difficulties merely technical?

L: Some problems technically could be very tough. Other problems you simply don't know how to approach. In that kind of situation, one has to be very original and have deep insight into the problem.

I: It seems there is a tendency to resort to computational methods when analytical solutions seem to be beyond reach. Is this a new paradigm in applied mathematics? Has this approach yielded new insights or breakthroughs?

L: In some way, I could say "yes". Historically, computation is an auxiliary tool. When we have difficulties understanding something, let's do some computation or we do some computations to verify. So therefore computation is always something supplementary to facilitate certain ideas and prove they work or do not work. With the development of science and technology, the situation is changing drastically, particularly over the last 10 or 20 years. The use of supercomputers in modeling is not only to understand certain issues and to do computations or numerical simulation; it is also becoming a preliminary kind of science. For example, in earlier times, we do a lot of experiments in materials. Then from the experiments we propose some empirical model according to our physical intuition and theory. From the equations there could be something new. Then you test the theory with the experimental data and then you modify and add in more parameters and so on. This is the classical way of doing things. But now it seems you don't go through the theoretical part so clearly at the preliminary stage. Some simply feed into the computer various parameters, effectively hundreds of experiments, at the same time. So you have much more data collected and how to handle these data and from these data get a reasonable mathematical model has become a sensitive issue in itself. The use of computers to model has become a necessity. It is not necessarily separated... I have this wonderful idea that I want to test using computers. No, you can also get the idea from the computational experiments.

I: So simulation will give rise to new ideas?

L: Right. Simulation itself will give insight into the problem as well as understanding of the problem. The computer can generate tons of data. Afterwards, you have to understand what these data are. You use statistical methods or other methods. With some models you can still do a testing. This process creates a lot of new mathematics. Some of the ideas that people have these days are from ancient times and they dropped out after Newton–Leibnitz's calculus worked so nicely and simply in ideal situations.

I: But Newton did not depend on simulation to come up with the calculus.

L: Mathematics does have a unique position in science. Sometimes it's indeed surprising that something that you somehow have purely from the imagination and logical deductions has to do with the real world. It may be because imagination is a part of the real world.

I: Is there any discovery of yours that you find intuitively surprising?

L: It's very hard to say. Sometimes when you prove something or create something you find it surprising. But after several years of deeper thinking and understanding, you realize that it's so natural. I find most of the things I did are indeed very natural. At a certain stage, something happens and one is surprised once in a while. For example, sometimes by looking at seemingly more complex problems, one can do much better. In the beginning it was a kind of surprise, but after years of thinking and understanding one realizes that it is very natural.

I: In your work on partial differential equations, do you put much emphasis on the beauty of the model rather than the technical details?

L: I'm personally much more interested in the ideas and methods that solve the problem. Sometimes there are certain technical computations you cannot avoid and you have to be able to handle such difficulties. Sometimes the technical things are the real things. But sometimes you are interested in the idea and the approach to the problem — which may be more beautiful and useful.

I: I believe some physicists believe that if a theory is beautiful, it must be correct.

L: In a certain way. If something is very simple, very beautiful, you say, "It's fantastic." Sometimes one can understand it from basic mathematics. But simple things could involve very deep and complicated mathematics. So you never know.

I: Is the Ginzburg–Landau equation completely solved in dimension 2?

L: There are a lot of papers and books on the Ginzburg–Landau equation in 2D and 3D. In a certain sense we understand quite a lot about these equations and their solutions. In partial differential equations, it seems you have so many equations to work on. But the good thing is that only a few equations are really fundamental and interesting. These equations will appear now and will appear later again and again. So I won't say that we completely understand the Ginzburg–Landau equations. It depends on what kind of questions asked and what kind of issues you want to address. For example, the Laplace equation — people worked on it for 200 or 300 years, we practically understand every aspect of it, but once in a while people will tell you something new about the equation. Because the Ginzburg–Landau equation is one of the fundamental equations to model basic physical phenomena, it is a nonlinear partial differential equation and I think it will appear again and again. Even for 2D, there are some issues we don't understand. So I won't say it's completely solved.

I: The Navier–Stokes equations are extremely difficult to solve. Is it due to the fact that it is difficult to formulate radically new concepts within the framework of classical physics?

L: They are probably one of the most fascinating partial differential systems I know of. I have personally spent some time thinking about them, but not really a lot because you get nowhere. We realize there are a lot of difficulties in understanding the issues but we don't know how to overcome these difficulties. Unlike a lot of mathematical questions, when you really understand the real difficulties of the problem you may try to find a way to overcome them. And sometimes you are lucky to solve them. For this one, looking at the difficulties from various views and angles, we understand it in a certain way but we don't know how to overcome them. Is it because of mathematically technical reasons or is it the formulation at the fundamental level? I really don't quite know. I won't be surprised maybe some day some people say, "This system is only one part of a grand physical system which may be solvable even though we don't understand this particular one." This may go back to the fundamental level of formulation of the problem. Maybe there is something missing from the very beginning. On the other hand, the Navier–Stokes equations are, from the mathematical point of view, already consistent and well-posed. In other words, it is a closed system and you don't need extra information from outside. But however, sometimes extra information from outside could lead to some fundamental or more radical ideas which may assist.

I: Is the existence part of the problem partly solved?

L: We understand existence under the so-called weak conditions, but we want a classical solution. People always tell you that it may not be important to real physics. But it is a very intriguing question, a simple mathematically formulated problem, and we simply do not know the answer. It's very mysterious.

I: But the physicists do not worry about the existence of solutions.

L: To the physicists, the physical system must exist but they may not talk about classical solutions. It's hard to say what one should really believe in. This is part of the difficulty.

I: The Navier–Stokes equations are classical and not quantum mechanical, isn't it?

L: Yes, there are many, many ways to derive that equation. Of course, from the mathematical point of view, you can forget whatever way you derive it.

I: Do the Navier–Stokes equations apply to all fluids?

L: Yes, but there are compressible or incompressible fluids, or visco-elastic fluids... One can derive similar equations in many real problems of physics.

I: Have they been extended or modified in some sense?

L: There are more complex forms of these equations and modified forms. But for the classical Navier–Stokes equations somehow the modified equations are couched in such a way that it is no longer interesting because the very difficulties of the original equations disappear. This is not the way to find a mathematical theory.

I: What are the chances of solving the equations in the next 30 years say?

L: It's very hard to make a prediction. Personally I don't like to make a prediction. But I would say it cannot be done in a relatively short period of time and may last a long time. For generations, the best trained minds have attempted the problem and it has defied all attempts.

I: It seems that applied mathematicians are generally more gregarious in research in the sense that they collaborate more among themselves than pure mathematicians do. Why is that so?

L: Yes, I also tend to believe that applied mathematicians are more gregarious than pure ones. I'm not surprised. But you also see more and more pure mathematicians joining efforts together to solve problems. It depends very much on the nature of the problem. Traditionally the mathematician works individually. But as the problems become more multidisciplinary and complex, it is natural to have groups of people working together to attack a common problem. In applied mathematics, the problems are by nature across the fields — it's mathematics applied to other sciences — and come from different disciplines. So it's not surprising. It should be this way.

I: What is your advice to the beginning graduate student who is interested in both pure mathematics and applied mathematics?

L: Just because you want to be cross-disciplinary, you try to understand a little bit from each field. That is, unfortunately, not the way. It is like that you are trying to understand the art better than they do. There is no reason you would do a better job than the experts from different fields; they can do much better than you do. Even if you do cross-disciplinary research, you have to be a specialist in one or two things — the insight, the ability to separate the problems. When you have that, you need to have an open mind, to learn things and get interested at the beginning. Even if you work in pure mathematics you shouldn't work alone in research. I may be more practical in a certain way but I'm interested in what is going on in pure mathematics and science and so on. If you have this kind of attitude and you specialize in one field, have an open mind in expanding your horizons to learn more things, you will do a very good job. People from different disciplines think in different ways. It's very interesting and good to know. Intellectually it's very satisfying. One would also find that there are many things in common.

I: Do you have any students?

L: At the moment I have four students — two are going to graduate, maybe next year. I have maybe 10 or 11 students who have already graduated. I also have some postdocs working with me.

I: Do you think that Chinese students are inclined more towards the applied side?

L: I don't think that way. They could be neither pure nor applied. The training in China generally seems to have certain disadvantages because students tend to focus and concentrate very early on one very special topic and remain so for most of the time. Once in a while you see a very top student but he is narrowly focused. If you are too focused very early on, then your ability will be very narrow also. If you don't expand your horizons and knowledge, then you lose your chance. Later on, when you see a problem you would say, "Oh, it's outside my field." Always take more topics courses. It shouldn't be that you take a course just to apply it to something. Of course, when you work on a problem, you will use whatever tools you have to solve it.

Pao Chuen

Of Science in Defense

Pao Chuen Lui (雷普权) played a pioneering and pivotal role in the establishment and development of the scientific and engineering capabilities of the Singapore Armed Forces (SAF) and the Ministry of Defense (Mindef).

Leaving behind his original dream of becoming a university lecturer after getting his bachelor's degree in physics from the then University of Singapore in 1965, he had a short stint at a radio and space research station before becoming one of the first scientists to join Mindef in 1966. Starting as a captain in the Logistics Division, he rose to the rank of colonel in 1978 in the course of a long and illustrious career in helping to build up, practically from scratch, the defense capability of a small nation that had to fend for itself economically, politically and militarily overnight. When he officially retired from SAF in 1986, he became the country's first Chief Defense Scientist. He went on in his typical tireless and visionary way to promote and coordinate research in science, engineering and technology in Mindef and SAF and to tap the resources of the universities and national research institutes in areas such as information and

Lui

signal processing, remote sensing, experimental hydrodynamic studies, high performance computing, computational mechanics and biomedical research.

In 1973, he obtained a master's degree in operations research and systems analysis from the US Naval Postgraduate School (NPS). While contributing much to defense science, he came back, in a full circle, to his original aim of teaching in the university when National University of Singapore appointed him as Adjunct Professor in Industrial and Systems Engineering in 1990. His contributions earned him professional recognition as Fellow of the Institution of Electrical Engineers (UK), Chartered Electrical Engineer of the UK Engineering Council and Senior Fellow of NPS. He was awarded the Commander of the Royal Order of the Polar Star by Sweden, NPS's Distinguished Alumni Award, induction into NPS Hall of Fame and the 2002 Singapore National Science and Technology Award. In addition to being appointed to leading positions in many national bodies such as Defense Science Organization, Defense Science and Technology Agency, National Research Foundation and Temasek Laboratories, he has been a member of the IMS Scientific Advisory Board that chartered the Institute's direction right from the beginning in 2000. Though he has recently relinquished his post of Chief Defense Scientist in 2008, he continues as a consultant to Mindef.

Lui was interviewed on 3 January 2007. The following is an edited and vetted version of the interview in which he gave us an insight into an unusual and distinguished career in defense science and how his scientific team questioned conventional wisdom encapsulated in the safety codes of the US Department of Defense and introduced new codes that are now scientifically accepted.

Imprints: You started schooling in Singapore at a rather late age after your family moved from China and Hong Kong. What is your most memorable impression of education (whether in school or university) in those days?

Pao Chuen Lui: I moved to Singapore in 1954 when I was eight years old. In school, I had to catch up with the rest of my cohort. I needed to move much faster and the school was very kind to me. My first school was Balestier Road Boys' School (for a year) and after a double promotion, I went to St Patrick's. The school was quite different in those days. In 1962, I was a boarder in St Patrick's and had a chance to be away from my family. We were boys of eight years old staying together in a very disciplined type of environment. Every morning, we attended classes and after studies, we played games in the afternoon. We had a lot of freedom and I enjoyed doing things in the Scout Movement. I also spent a lot of time in the library reading mathematics. When I entered university [University of Singapore] I had one clear objective — to graduate, get a scholarship and come back with a PhD. Unfortunately, I didn't do well enough in my final exams. I didn't get a First Class Honors and I could not get a scholarship to go to Cambridge. That was when I decided to start working.

I: You could have gone to some other university for graduate studies.

L: I set my mind to do something. When I couldn't get it, I would do something else.

I: What did you do after that?

L: The UK Science Research Council had a Radio and Space Research Station in Singapore. It had two sets of facilities, one at the Bukit Timah campus, next to the Students' Union House, with equipment to measure the ionosphere and the other was at the Sembawang Air Base where they were tracking the UK1 and UK2 satellites and gathering data from satellites passing over Singapore. I was employed by the UK Science Research Council to work at both facilities. After one year, I found that the Singapore Armed Forces had opportunities for me to contribute to Singapore, and so I switched.

I: Your initial training was in physics, but your career was more in engineering or at least technology. Did you ever consider doing an engineering degree in the first place?

L: Our family finances did not allow me to study engineering. At that time, you had to go to KL [Kuala Lumpur] to do engineering. The engineering faculty was in MU [University of Malaya]. At that time, SU [University of Singapore] was part of MU, and they [MU] did not have a medical faculty while SU did not have an engineering faculty. To do engineering, you had to go to KL or do it at the polytechnic [Singapore Polytechnic], but it was only a diploma course, not a degree course. Reason number two is that engineering was a four-year course, whereas for physics, you could do it in three years by going directly into the second year as a "super-fresh" student [on the basis of good 'A'-level grades].

I: After you were appointed as the Chief Defense Scientist in 1986, your work seems to move more towards research capabilities in science and technology.

L: Before that, I was running special projects. My last assignment was as a Project Director for Mindef and SAF. One major capability development is in science and technology capability in SAF and Mindef and to develop international relations.

I: Do the research capabilities have to be connected to defense?

L: Most of the capabilities we do must add value to our defense capability, including the R&D [research and development].

I: What is your latest research initiative?

L: I'm more involved in the management of research and development rather than doing research myself. For example, we have a problem of space in storing a large amount of ammunition in Singapore. Can you find a way where you require less land? There is so much land required if you store it on the surface. If you were to go underground, below the rocks, the current codes say that you require 240 hectares. But we looked deeper into it and found that actually those codes were not based on science but on empirical considerations. We did research and development on the ground shock behavior of explosives when they explode and we came out with a code that requires much less land than that. We embarked on a very large technology development program. Firstly, you must understand the physics, do the modeling of how explosives detonate and study their detonation patterns. Originally, we assume that explosives detonate as point masses, but we realize that's not the case in real life.

They cannot be point masses, they are distributed in storage. When they explode in storage, they can go off together provided they detonate together. But would the peak pressure be the same as a point mass? We then did some computer simulations and found that the pressure is very slow if you assume that they are distributed and you have a very simple assumption that they detonate at the same time, which is a contradictory type of assumption. So then, what else are there? We know that the ammunition is axially distributed. We looked at a number of critical parameters. First, the ground shock: will the buildings nearby collapse? You also need to know the structural behavior through soil, through layers of rock and so on. We thought that there must have been some research done in those things. The very interesting thing is that in the US Department of Defense, for safety codes, there were two formulas used to determine ground shock. So we put in the amount of explosives that we intended to store into the formula. One formula came up with an answer of 60 hectares and our formula came up with something else. We went back to the Department of Defense and there was a big meeting. We had two sets of formulas and two different answers; so one of them must be wrong. Unless we have the scientific basis, we cannot persuade you to accept our findings. So we went back to the mathematical formulation, computer simulation, small scale testing and finally large scale testing. We built a tunnel in Sweden to verify our model. After that, the results were presented to the experts in NATO. They looked at our results and found that there was a scientific basis for us to write a new code based on our findings. What does that mean?

The normal way of storing ammunition above ground requires 1,000 hectares of land. If you were to go below the rocks, in the caverns, making use of the present safety codes, the land required is 240 hectares. We found that these are empirical codes without any scientific basis to it. We went back to the scientific basis of the effects of the explosions — the theory, modeling, small-scale testing and large-scale testing — and found that the amount of land required is 100 hectares. This new code is now embodied into the safety codes of the US and NATO.

I: Is it implemented in Singapore?

L: Yes, of course. Why would we want to spend S$12 million on R&D to find that out?

I: In some countries, there is a perception among some academics that research sponsored by defense agencies tends to be geared towards military objectives and therefore not "peaceful" in nature. What is your view about this?

L: In Singapore, it is very clear. If there is no defense, there will be no Singapore. The economic well being of the country depends on the perception of investors that Singapore is a safe place to invest their money in. In order to convince investors that Singapore is a safe place, national security must be given top priority by the government. With national security comes development, next economic development, then social development. National security is fundamental to the well being of Singapore. NUS defense technology is a strategic thrust of the university and is important for our defense.

I: Has NUS been given any grants by Mindef?

L: Mindef identifies some research which can be done by researchers in the universities and accordingly funds the researchers to do this research. In order for results to be able to be fed into DSO [Defense Science Organization] they must create systems that are of high mutual value.

I: Does Mindef approach individual academics to do the research?

L: It started more than 10 years ago with road shows that we conduct every year with briefings of what we are doing to academics to see whether they have any interest. Therefore it is a two-way thing.

I: Could you tell us what is defense systems engineering? What is the role of mathematics in this discipline?

L: Systems engineering ensures that all the components of an enterprise are blocked together in such a way that they will achieve the objective of that enterprise, by looking at the topologies and components of the subsystems, and in particular, defense systems engineering as it pertains to defense systems. The role of mathematics is fundamental. In anything that we do, it provides the tools for a scientific basis.

I: Is any of the research done by scientists in DSO considered to be "classified" and hence not published in the public domain? If so, would that be contrary to the spirit of free inquiry that is so dearly cherished by creative scientists?

L: There is protected work done by DSO. One is the development of capability that is normally systems specific and that is classified. They are creating capability that is of high value and that you cannot buy. These are the things that are classified and we will not tell people what we are doing until we know that the value of that information is no more relevant. But in the case of research, like lasers and mathematics, it is publishable. Some of the researchers are adjunct professors in the universities. There are research intensive type of publications. There is work on developing capabilities and we don't publish them. There are two parts of DSO: the open part dealing with more fundamental and basic research, and the second systems part of DSO.

I: Do you have people working on both types?

L: Yes, they can work on both.

I: Has "biological warfare" been seriously considered a potentially lethal option to be countered in defense science and are there any effective measures against such options?

L: More than 10 years ago, we established a biomedical initiative and one of the areas of focus was to study how biological agents could be spread in a densely populated place like Singapore and how we can respond. Biological activity has been a threat and continues to be a threat. Therefore the defense, medical and environmental research institutes play a part in total defense and have a responsibility to do research in the solution of this problem.

I: Is there any collaboration between DSO and biological scientists?

L: Our capacity is very limited. That is the reason why the billing in the BMRI [Biomedical Research Institute] in the medical faculty is for scientists in DSO to clap hands with the medical faculty.

I: Is there some kind of biological initiative initiated by DSO or Mindef?

L: I wouldn't say biological initiative. Perhaps research in defenses against biological agents.

I: You mentioned that there are people in DSO who do more publishable research. Are they free to choose their kind of research problems?

L: Most of their research is in common areas of interest rather than in the areas of interest of the investigators. The investigators know the ultimate areas of interest to defense science and therefore, if they want to publish in astrophysics say, then probably they are not going to get much financial support from Mindef or DSO.

I: What are some of the "pure" areas relevant to DSO?

L: Electromagnetics. There was one book on electromagnetics. What I mean is information theory — or rather information security, especially with computer systems. We have a number of mathematicians working on how to protect information and how to understand information that could be extracted from noise.

I: Are computer scientists involved?

L: We have a sizable group of computer scientists working in defense.

I: Are there many of our graduates working in DSO?

L: One thousand something.

I: What are some of the achievements that you would like to be remembered for?

L: The things that are very interesting I can't tell you because they are classified, and the things that I can tell you are quite mundane. One thing I would like to be associated with is the build up of the core of engineering capability in SAF. When I joined the Logistics Division, there were only a handful of us with some science background. Now we have up to 3,000 science and engineering graduates. It takes a considerable effort to build, equip and train first rate engineers and then develop them into an organization. I'm quite proud to be associated with the people and organization that will be able to provide that depth in science and technology that allows them to have the confidence in pressing beyond world class caliber.

I: Do you think of retiring?

L: I'll be retiring in March 2008.

I: In spite of your heavy administrative involvement in many companies and committees, you have found time to teach in the engineering faculty of NUS for about 10 years. What is the greatest satisfaction that this has given you?

L: More than 10 years. I started teaching in 1990 because the engineering faculty thought a minor in development systems would add value to the faculty. I teach one module a year on large scale systems engineering. I feel that it is important for the older generation to share their knowledge with the younger generation through teaching. The way I teach is through story-telling and case studies and to relate it back to some of the theory that are all there. I find that when the students meet me after 10 years, they still remember. It's not the theory but the stories and the lessons learned.

Eric

Game Theory Master

Eric Maskin has made fundamental and groundbreaking contributions to numerous areas of game theory and economic theory, such as implementation theory, auction theory, the economics of incentives, and social choice theory. Among the wide range of topics he is currently studying are the design of auctions, comparison of different electoral rules, the pros and cons of political accountability, and the advantages and drawbacks of protecting intellectual property. His work, sometimes in collaboration with co-authors including Partha Dasgupta, John Riley, Jean Tirole, Jean-Jacques Laffont, Peter Diamond, and Drew Fudenberg, has been widely applied to industrial organization, finance, development, and other fields within economics and political science.

He has published over 100 articles and book chapters, and has served on the editorial boards of leading economics journals such as the *Quarterly Journal of Economics*, *Economics Letters*, *Social Choice and Welfare*, *Games and Economic Behavior*, the *Review of Economic Design*, the *Review of Economic Studies*, and the *Journal of Economic Perspectives*.

Maskin

He has been frequently invited to give named lectures, in particular, the Arrow Lectures, the Mckenzie Lecture, the Zeuthen Lectures, the Schwartz Lecture, the Marshall Lectures, and the Pareto Lecture. He is a Fellow of the American Academy of Arts and Sciences and of the Econometric Society, a Corresponding Fellow of the British Academy, and an Honorary Fellow of St John's College, Cambridge. He was President of the Econometric Society in 2003.

He was a Research Fellow at Cambridge University (1976–1977) and taught at MIT (1977–1984) and Harvard University (1985–2000), where he was Louis Berkman Professor of Economics (1997–2000). He then moved to the Institute for Advanced Study at Princeton and is currently Albert O. Hirschman Professor of Social Science. In 2007, he was awarded the Nobel Prize in economic sciences together with Leonid Hurwicz and Roger Myerson.

He was interviewed on 16 May 2005 when he was at IMS as an invited speaker of the program on uncertainty and information in economics. The following is an edited account of the interview in which Maskin talked passionately about the revolutionary game-theoretic ideas that are changing economic theory and influencing the social and political sciences in practice.

Imprints: Your doctorate at Harvard was in applied mathematics. What kind of applied mathematics was it? Was it related to economics?

Eric Maskin: In those days, applied math at Harvard at the graduate level was a fairly free-form program. Each student designed his own program of study, and the only common requirement was that the dissertation had to have significant mathematical content. My own program included a fair amount of economics, and, in fact, my advisor was the economist Kenneth Arrow [Nobel Laureate 1972].

I: You were in the Mathematics Department?

M: No, I was in Applied Math. It wasn't a "department" *per se*, but an interdisciplinary committee, including some people from the Math Department, a few from the Economics Department (in particular, Arrow), and assorted others from Statistics, Engineering, and so on. I did quite a bit of mathematical economics. I also did some mathematics not related to economics simply because I was interested in it. My thesis was in game theory and social choice theory.

I: You went to Cambridge University immediately after your doctorate. Was there any specific reason for that?

M: While I was studying at Harvard, the Cambridge economist Frank Hahn — a close friend and collaborator of Arrow — visited for several months. Hahn encouraged me to spend some time in Cambridge in a postdoctoral position. The idea appealed to me: Hahn was a leading mathematical economist and so it made sense educationally; I also thought it would also be interesting culturally.

I: Did you do any joint work with Hahn?

M: I never wrote a joint paper with him, but talked to him a great deal about my own work — he was extremely generous about making time for that. I've remained personally very close to him since those days. In fact, I'm going to England at the end of this week to help celebrate his 80th birthday.

I: You taught at MIT for a short period after returning to the United States from England. Is it true that one of your most important works was done during this period? Could you tell us something about it?

M: Yes. Actually, I wasn't at MIT for such a short a period; it was seven years. I worked on many things there, but probably the thing that I'm best known for now was a series of papers on implementation theory. In implementation theory, the idea is to construct a game or mechanism for attaining the goals that you as the mechanism designer wish to achieve. Suppose, for example, that there are some economic resources to be distributed among the agents in the economy and that you, the designer, have a particular set of criteria in mind for evaluating different possible distributions. Imagine however, that determining the best distribution according to these criteria calls for information that you don't have. Assume that the agents in the economy have this information, but you don't. Then, what you can try to do is design a game for the agents to play so that, when equilibrium is reached, the outcome is the same as the one you would have imposed had you had that information in the first place. In effect, the game itself "compiles" the agents' information, enabling the right outcome to emerge in spite of your own ignorance.

I: It sounds a bit psychic.

M: Well, it's what economic systems are supposed to do. An important reason why market economies have historically worked better than planned economies is that typically economic planners don't have enough information to allocate resources appropriately, even if their intentions are benevolent. Markets, by contrast, provide a remarkably good way of aggregating or gathering information. Implementation theory can be thought of as a generalization of the sort of things markets do.

I: Is implementation theory part of game theory?

M: Yes, it is part of game theory. Implementation theory actually intersects a number of areas of economics, but because it presumes that people act strategically, it is certainly game-theoretic. Still, it differs from much other game theory in an important respect. Usually in game theory, we start with the game to be analyzed and then try to predict what will happen, i.e., what the outcome will be. In implementation theory, the process is reversed. We start with the outcome that we want and then consider how we might construct a game that achieves that outcome. So implementation theory can be thought of as the "reverse engineering" part of game theory.

I: You went back to teach at Harvard after MIT. Was it some kind of loyalty that made you return to Harvard?

M: It was partly loyalty. Also I was intellectually close to a number of people at Harvard, which had been my first academic home. Still, MIT was a terrific place to be. I enjoyed and profited from my time there enormously.

I: You once mentioned that game theory has revolutionized many fields, especially economics. Could you give us some examples of this in economics?

M: Yes. Before game theory, most of economics dealt with models of large markets — so-called "perfectly competitive" models, where there are lots of economic agents: lots of buyers, lots of sellers. Paradoxical as it may seem, the large numbers actually made the analysis easier, because the agents in those models didn't have to act strategically. If you are only one seller among many, you are not going to have much influence on the other sellers' outcomes, and so won't affect their behavior much. This means that you don't have to take into account how they are going to respond when you decide what price to set or how much to sell; your decision is relatively straightforward. Game theory provided economists with the tools for analyzing the small numbers case, which is the relevant case for many industries. In the US automobile industry, for instance, each of General Motors, Ford, and Chrysler is big enough so that, whenever it acts, it has to calculate how that action is going to affect its competitors and how those competitors are going to react. For economists interested in the automobile industry, calculating what is going to happen is therefore harder than predicting what would happen in an industry where there are lots and lots of sellers. Game theory helps us to make those calculations. It's been developed precisely to deal with the case where each player has a significant influence on the payoffs of the other players.

I: Someone once said that every economist should know auction theory. How much do you agree with it?

M: These days auctions are highly visible economic institutions. They are used in settings ranging from online sales of everyday items to the privatization of major public assets. So, one reason economists should know some auction theory is that auctions constitute an ubiquitous practical way of allocating resources. They also provide us with an explicit mechanism of price determination. How prices actually get set is an issue that is left out of many economic models. Auction theory gives economists the foundation for a theory of price formation.

I: Does auction theory apply to any number of participants?

M: Oh yes, auction theory is all-inclusive in that sense. But the theory is probably most interesting when there are relatively few participants because that is when the game-theoretic aspects come to the fore.

I: Is auction theory a species of game theory?

M: Auction theory can definitely be thought of as part of game theory; an auction is a game in which the bidders are the competing players. If I'm a bidder in an auction and considering what bid to make, I have to take into account how the other bidders might behave. Similarly, they will be thinking about what I'm going to do. So there is a strategic interaction that calls for game-theoretic analysis.

I: Have you personally organized any auctions?

M: Well, I tend not to get too involved in practical consulting, but I have made a few exceptions. One was a couple of years ago. The British government was interested in creating an auction to help reduce carbon dioxide emissions by British companies. The problem raised some interesting theoretical questions, and so I agreed to help design the auction.

I: Did they raise more money than they expected to?

M: They weren't actually trying to raise money in this case; they were trying to spend it effectively. The government budgeted about 300 million pounds (or about 500 million US dollars) that they were prepared to spend to induce British firms to decrease their CO_2 output. The question they put to us auction designers was how to construct an auction in which companies would be awarded money in exchange for pledging reductions so that this budget would stretch as far as possible, that is, the biggest reduction in greenhouse-gas pollution would result.

I: I understand that you have been recently applying game theory to the study of electoral procedures within a democratic system. Are there any clear-cut answers?

M: In fact, that work, which is joint with Partha Dasgupta, is not entirely game-theoretic. The issue we examine is whether there exist voting procedures that satisfy certain basic desirable properties. A famous theorem obtained by my thesis advisor Kenneth Arrow over 50 years ago establishes that there is no voting procedure that satisfies all these properties all the time. Thus, the natural question to ask is: Which voting procedures satisfy the properties in the largest class of cases? It turns out that there is indeed a sharp answer.

Specifically, the procedure called "simple majority rule" (or "Condorcet's method," after the 18th-century scholar, the Marquis de Condorcet) is the voting rule that satisfies these properties more often than any other. Of course, I have to tell you what these basic properties are. One, called the "Pareto (or consensus) principle" says that if everybody in society prefers candidate X to candidate Y, then candidate Y should not be elected. Another property requires that all voters should count equally. It's sometimes called the "anonymity" or "one-person, one-vote" principle. A third property, "neutrality," has two components. The first is symmetry, which means that the electoral rules should not favor one candidate over another. The second requires that the voters' choice between candidates X and Y should not depend on their views about some third candidate Z. The final property, called "transitivity", demands that if candidate X is chosen over Y, and Y is chosen over Z, then X should be chosen over Z. From Arrow's theorem, there is no voting procedure that satisfies these four principles all the time. But simple majority rule satisfies them more often than any other rule.

Simple majority rule compares candidates pairwise. If there are three candidates X, Y, and Z running, we should elect candidate X provided that X would defeat Y in a head-to-head contest and also beat Z in such contest. Of course, in most elections, you can't vote for more than one candidate at a time, which means that it may be impossible to tell from the returns alone whether X would beat both Y and Z in this way. Therefore, Dasgupta and I recommend that voters should be allowed to express their rankings of all three candidates. Rather than simply voting for candidate X, you might, for example, write X first, then Z, then Y. When all voters supply these rankings, one can make the pairwise comparisons that I was just talking about.

I: Is it possible to get a situation where there is no clear-cut winner, that is, no candidate who will beat the other two?

M: That is possible. In fact, that possibility was discovered by Condorcet himself. In such a case, one would have to use some tie-breaking rule to determine the winner. Still, even though simple majority rule doesn't always work as it is supposed to, it works, as I was suggesting, more often than any other voting rule. The cases in which there is no clear-cut winner turn out to be less numerous than the instances in which other voting rules run into trouble. Thus, there is a sense in which the problem you point out is less serious than those that beset other voting rules.

I: Has this method been tried in out in practice?

M: I don't believe it has been used in a large-scale election, say on the national level. But it has certainly been used in smaller elections, e.g., those for committee officers. Now that it's so easy to vote and count votes electronically, I think the time has come to try the method out on a bigger scale. The current way of determining winners in Congressional and Parliamentary elections in the United States and the United Kingdom — the winner is the candidate with the most votes even if that total is short of a majority — is, in my view, highly objectionable. With more than two candidates, it's quite possible for the winner to be a "minority" candidate in the sense that a majority of voters prefer some other candidate. I would like to see simple majority rule used instead in these elections.

I: You have also done some work on social choice theory. Could you tell us briefly what this is?

M: In a sense, we have already been talking about social choice theory: it includes implementation theory and voting theory. Social choice theory examines the question: How do we go from individual preferences to social preferences? Of course, we do that when we vote, but the theory goes well beyond voting. Almost any public policy question involves passing from individual to social preferences. When we decide how much public education to provide, how much to spend on national defense, or whether to redistribute income from rich to poor, we have to answer that question. Thus, all these issues belong to social choice theory.

I: These are very practical issues.

M: They certainly are. The theory itself is generally worked out at a fairly abstract level, but it often has strong implications for practical issues.

I: Is it easy to convince the politicians?

M: I never try [*laughs*]. I wouldn't know how to begin to convince politicians.

I: In some of your papers, you drew some analogies between economic behavior and animal behavior with respect to evolutionary biology. Could we interpret this to mean that animals are also "economic" creatures?

M: If by "economic" creatures, you mean creatures that "optimize", I think they definitely are. In order to survive, an animal needs (i) food, (ii) shelter from the elements, and (iii) some way of dealing with other animals that might harm it. How an animal pursues these objectives is its "strategy". Evolution serves to select animals with successful strategies over those with strategies that don't work so well. The animals with the successful, "optimizing" strategies will survive. They will have the opportunity to reproduce, and their offspring will carry on the optimizing legacy. So, evolution forces animals to be optimizers in the same sense that consumers and producers in standard economic models are optimizers.

I: Have you tried to apply your ideas to evolutionary biology systems?

M: Absolutely. The paper I gave today at the conference — another project with my old friend Partha Dasgupta — is on this exact topic. We are trying to understand a certain kind of perplexing behavior that has been observed in certain bird species and also perhaps in humans. Humans are considerably more difficult to study because their behavior is so complicated; it's usually harder to do good laboratory experiments with human subjects than with pigeons. Anyway, the paper constructs an evolutionary model that attempts to account for the documented behavior.

I: A related question: does this mean that economic behavior may have genetic origins?

M: Undoubtedly, many aspects of our daily behavior do have genetic origins. The fact that we are impelled to eat when our stomachs are empty is programmed into our genes. But more complicated optimizing behavior (such as the way we invest in the stock market) is often so complex that usually one can't say how much of it, if any, is genetic in origin. That's why evolutionary psychology is such a controversial subject; in the case of humans, it is very difficult to disentangle what is the result of biology, what is the result of rational calculation, and what is the result of past experience.

I: It seems that economics is now so intertwined with so many other fields.

M: That's certainly true. In particular, the boundary with psychology is where some of the most exciting work in economics is being done today. But the boundaries with other social sciences — especially political science — are also very lively.

I: Is the urge to take risk genetically driven?

M: I don't know the answer to that. There are those who suggest that there may be a genetic disposition toward taking risks. It has been conjectured that males may be more disposed to taking certain kinds of risks than females. That's not a question I have looked into myself, but it's fascinating.

I: Economics is so different from what it was 20 years ago.

M: Yes, it's a subject that has evolved quite rapidly. That's one reason why it's stimulating to be an economist these days — the subject changes so quickly.

I: Do you do simulation in your type of work?

M: I don't personally use simulations often, not because I'm opposed to them but because I prefer, as a matter of taste, doing things the old-fashioned way by calculation. Simulation can be a useful first step for formulating and ruling out conjectures but, in my view, it should rarely be a last step. After you have done enough simulations to have a reasonably good intuition about what is true, then I think you should sit down and try to prove things analytically.

I: In auction theory, for example, there should be some scope for simulation.

M: Sure. In fact, often before a new sort of auction is used in practice, the designers will try to simulate bidding to make sure something won't happen that they didn't think of. Simulation can be a powerful short-cut. I just don't think it should take the place of theorem-proving.

I: Was there any specific reason why you moved from Harvard to the Institute for Advanced Study at Princeton?

M: Yes, there were a couple of reasons. Harvard was a fantastic place to be, and it has as good an Economics Department as one could hope for. But one advantage I have at the Institute is unparalleled freedom — freedom that a professorship in an Economics Department could never bring me. At the Institute I have few formal duties and the time to do pretty much whatever I want. The other thing I get there is interdisciplinary contact. In the School of Social Science (where I am), I bump up against anthropologists, political scientists, historians, etc. all the time. I find this interdisciplinary mix stimulating and interesting. I think one of the ironies about modern universities like Harvard is that one has an enormous variety of interesting people nearby, but little opportunity to talk to them. If you are an economist, you see the economists, of course, but you don't run across people in other fields nearly as often. One nice feature of the Institute is that you talk to non-economists every day, and so and it's not an extraordinary event to have a truly interdisciplinary conversation.

I: Are there other economists in the School of Social Science?

M: I'm the only economist on the faculty, but we always have economists visiting. At the moment there are five economists at the Institute for the year, and so I certainly don't feel lonely.

I: Do you have any students at the doctoral level?

M: I do, I have always had PhD students. I think of that as an essential part of my professional life. It's not only fun supervising these students, but useful for my research. Advanced students, particularly those working in areas close to mine, often have stimulating new ideas, which cause me to rethink what I'm doing. In some cases, these students turn out to be collaborators. Some of my favorite co-authors are former students.

Eduardo

Infectious Diseases, Vaccines, Models

Eduardo Massad is internationally acclaimed for his work in the field of infectious disease epidemiology, both in the mathematical modeling of infectious diseases and in the introduction of successful disease control programs in Brazil.

Originally trained in medicine, Massad went on to obtain a degree in physics and a PhD in experimental pathology. Together with his collaborators, he did important work in the mathematical modeling of infectious diseases such as dengue, yellow fever, hepatitis A, vaccine preventable diseases, parasitology, HIV and antimicrobial resistance.

His extensive research work is contained in almost 200 joint papers and four books. During his distinguished research career, he has supervised 21 doctoral students. His scientific contributions extend beyond the theoretical aspects of modeling in infectious diseases and epidemiology to the successful implementation of disease control programs in Brazil. Fluent in six European languages, he also has research interests in evolutionary biology and telemedicine.

Massad

Massad is Professor of Medical Informatics at the University of São Paolo in Brazil and has been an Honorary Professor of Infectious and Tropical Diseases at the London School of Hygiene and Tropical Medicine since 2003. He is a Chartered Mathematician and Fellow of the Institute of Mathematics and Applications and a Chartered Scientist, Science Council, United Kingdom.

In 2003, the Courage Fund was set up in Singapore to meet the deadly challenge posed by the Severe Acute Respiratory Syndrome (SARS), and a grant of S$3 million was presented to NUS to set up the Visiting Professorship/Fellowship in Infectious Disease and Epidemiolgy. In 2005, Massad was invited to Singapore as the inaugural Visiting Professor to initiate the building up of expertise in epidemiology management and control. In 2007, he was Chair of the Institute's *Workshop on Mathematical Models for the Study of the Infection Dynamics of Emergent and Re-Emergent Diseases in Human* held from 22–26 October 2007. During his visit, he was interviewed on 29 October 2007. The following is an edited transcript of the interview in which he gave us a rare insight into the collective efforts of medical and academic specialists in the ongoing prevention and control of infectious diseases around the world. We also get a glimpse of how mathematics, statistics and computer science have put in their share of efforts in the continuous struggle to maintain the healthcare of humanity.

Imprints: You originally started with a diploma in medicine and then went on to concurrently obtain a bachelor's degree in physics and a medical degree. Why did you choose a degree in physics, why not chemistry, say, or even mathematics?

Eduardo Massad: Well, it's actually not that easy to describe. I did not choose chemistry because chemistry was not exactly what I wanted to do. I was in doubt between physics and mathematics. I chose physics because that course at my university at São Paolo was more applied than the mathematical one. I intended to be an applied scientist. I then chose the physics course degree because there was a balance of applications in that course.

I: Was your education taken mainly in South America?

M: Yes. I was trained in São Paolo and then I did my graduate study in England.

I: When you say graduate study, are you referring to…?

M: I mean, as a postdoctoral graduate in England.

I: Which part of England?

M: I stayed for one year in Sussex University, and then for two years at Imperial College.

I: Was the training in physics directly useful in your subsequent research work in the medical sciences?

M: Actually, it was not really useful; it was determinant because everything I do is professional nowadays. Thanks to my training in physics, it was very central indeed.

I: So it's the training rather than the actual knowledge?

M: Yes, the form of training.

I: One of your research areas listed is "medical informatics". How different is it from bioinformatics?

M: Well, it's quite different because actually bioinformatics is part of medical informatics in a sense. Bioinformatics deals strictly with information from genomic science and the huge amount of data that is coming from the Human Genome Project while medical informatics deals with information in medicine as a whole, with medical records, artificial intelligence as applied to medicine and so on.

I: It's a much wider science.

M: Much wider than bioinformatics.

I: Is this a very new kind of discipline?

M: Bioinformatics is new but medical bioinformatics has a history of 35 years.

I: Did it come about as a result of the computer revolution?

M: It's a compromise between computer science and medical doctors in the sense that there is a growing interest in applications of computer science to solve medical information problems.

I: In the early years, they were talking about expert systems.

M: Yes, expert systems is one of the specialties of medical informatics. There are a number of very interesting systems that are able to do medical diagnosis nowadays.

I: How recent is the field of mathematical modeling in the medical sciences?

M: In the sexual disease field, it's quite old in the sense that it was the first field to which mathematical thinking was applied in order to understand events in sexual diseases... Sir Ronald Ross in 1911, 1918 — he was the man who discovered that malaria was disseminated by mosquitoes. Then in the '20s, a couple of researchers started to apply more formally mathematics into the study of sexual diseases, and in the '70s, the field exploded. So it's about 40 years and it has consolidated since then.

I: Is it true that not many doctors are well equipped to do modeling?

M: No, not that many doctors. You see, it's very rare to have a doctor apply mathematics directly.

I: You are involved in training doctors to do the modeling?

M: Yes, yes. I have many doctor students.

I: Are there any general laws or principles that can be formulated in medical modeling like those in physics or chemistry?

M: Oh yes, certainly there are. There are some central principles that you can call general laws, like the law of mass action that is interaction between susceptible and infected people, the basic reproductive number.

I: Do epidemiological models depend on the size of the domain affected? In other words, are there such things as local models versus global models?

M: Very much so, yes. You have global models that are applied to explain diseases anywhere in the world. You have global models for certain kind of infection like vector-borne disease, and you have local models that are applied to specific communities, whereas the global models can be applied to any community whatsoever. You can distinguish between local and global models. There are certain models that are applicable only to certain communities.

I: How successful is computer simulation in predicting or preventing disease outbreaks?

M: I think it is very successful indeed. The problem is to convince the decision makers to believe in such a model. So the main problem in the actual application of such predictions is convincing the decision makers that the model is correct and reliable. The main problem is communication rather than the technology.

I: How do you convince the policy makers?

M: Normally, the logic behind some models can be explained to them, but if you make them too complicated, they tend not to believe you.

I: I think most of them would believe you if you can show them the results.

M: Yes.

I: I believe you have actually applied some of these models to solve problems in São Paolo.

M: Yes, we applied the models to real problems there.

I: You have no problems in convincing the authorities there?

M: Well, some 15 years ago, we had one Secretary of State in São Paolo who really understood what our model was about and he was the one who supported the application of models as a whole in the state of São Paolo. We were lucky to find this man behind the job.

I: Have you applied your methods to formulate ecological models?

M: Yes, indeed. We applied some of our methods to simulate ecological models for vector-borne infections. The ecology of mosquitoes in the disease can be formulated according to the mathematical approach.

I: In your models, you have some kind of parameters where you have to measure something. How do you decide what these parameters are?

M: The parameters are determined by the model you structure. In a sense, the theoretical model anticipates the field work.

I: Before you formulate the model, do you collect the data first?

M: Yes. You know the problem and you know the natural history of the disease. Then you formulate a model that mimics this natural history. Then you go after the parameters in the field in order to check the model's position.

I: Does that mean you have to refine the model again and again?

M: You refine the model according to the data.

I: How do you know when you have reached the ideal model?

M: Normally if you know the natural history of the disease well, you don't have too many competitive models to choose from. You normally have only one model that mimics the natural history. The structure of the disease must be behind the structure of the model.

I: Are your models stochastic or deterministic models?

M: They can be either stochastic or deterministic. It depends essentially on the size of the population involved. For smaller populations, you have to apply stochastic models. For larger populations, you cannot apply a simply deterministic approach.

I: How do you pick up the necessary mathematical tools?

M: I was trained as a physicist. The essentials I know, but I work together with more accomplished mathematicians who help me on the project.

I: Are your models intuitively motivated?

M: Oh yes, intuition is very important indeed and is an essential part of the job.

I: Does that come from experience?

M: It can come from experience.

I: You have been highly successful in using modeling to control disease outbreaks in São Paolo. What was the greatest breakthrough behind those successes?

M: I think the greatest breakthrough is that we manage to succeed with the model which determines the introduction of the rubella vaccine in the state of São Paolo in 1992. At that time, the health authorities have some doubts whether they should introduce the vaccine or not because, you know, rubella is a problem that affects pregnant women. And the vaccine causes a shift in the average age of infection. Before the introduction of the vaccine, the average age of infection for rubella is about six years of age. But when we start to vaccinate, this age shifts about. The fear at that time was that they did not manage to vaccinate the maximum portion of children in order to avoid the reproductive window which is between 15 and 40 years of age. We manage to calculate that and design a model which guided the introduction of vaccine in the form of a mass vaccination campaign. After the mass vaccination campaign, we introduced the vaccine in the local calendar of vaccination.

I: Was the result immediate?

M: Well, the immediate result was that the number of congenital rubella syndrome, which is something that affects the babies, dropped close to zero one or two years after the introduction of vaccination.

I: That is rather dramatic.

M: Yes, very dramatic indeed.

I: You collaborate with a lot of people. It's typical in biology, isn't it?

M: Yes, in biology it's common. Biology is team work. In the medical profession, even if you are working with some basic side of a medical problem, you have to work with a team of people in the background. We have doctors trained in infectious diseases, epidemiologists; we have mathematicians, statisticians, psychologists, meteorologists — a lot of different people working together.

I: Are diseases dependent on the actions of human beings?

M: Oh yes. Almost any infectious disease is behavioral dependent. It depends on how people behave in order to get the disease. It's important to know that.

I: How do you control the behavior of people?

M: You can't control the behavior of people. You just can only educate people.

I: What about intangible things like mindset and cultural attitudes?

M: That's the basic barrier in the HIV epidemic. It's a clear example of that — you know, to make educational contact in order to inform people what are the risks and the behavior that they should follow or what they should avoid and so on.

I: You have many research students. What is the secret of your ability to attract so many students into your research area?

M: Well, first it's because they were available. You know, full-time research in the medical school is a rarity. The sheer availability attracted some students. Second aspect is the novelty of this kind of method that I tried to apply some 25 years ago. So this attracted some students mainly because of their curiosity and the novelty of the approach. But, more recently, I'm not attracting so many students any more, mainly because I have to compete with microbiology and other novelties at this time.

I: These are the newer areas.

M: Yes, newcomers, and you have to compete with young and bright people with other interests.

I: What are your views about the prospects of controlling diseases in the future?

M: I think that vaccination is the key for any infectious disease. As soon as you find an effective and cheap vaccine, you can control any kind of infectious disease. The problem is that for some diseases, it is more difficult to develop vaccines, like dengue fever.

I: Vaccination is more like prevention. But once a disease has started...

M: Once a disease has started, you have to treat it. That is the greatest challenge of modern medicine — to find new antimicrobicide, antibiotics, anti-viral drugs.

I: What about global warming issues? Do you consider them?

M: Yes. At the moment, we are looking at one model that is trying to understand how the seriousness of diseases can be affected by global warming; in particular, those diseases that are transmitted by mosquitoes.

I: What is the impact of the Human Genome Project on diseases?

M: The Human Genome Project has a huge impact on the understanding of diseases — the fact that it can pin down the genes that are responsible for specific susceptibility to certain infection. Other parameters that are important are those genes that may facilitate the entrance of micro-organisms into our bodies and how they penetrate into our cells, how our organisms interact with this sort of micro-organisms, what is the best treatment with drugs in that specific genomic setting. Well, I would expect a huge impact.

I: What would be the ethical issues involved?

M: That is a very sensitive point because the ethical conscience would find new barriers for some kind of investigations. For instance, investigations with animal models are very much restricted nowadays for ethical reasons. And also observations on experimental human beings are very much constrained by ethical issues.

I: You mentioned animal models. Is animal testing done anymore?

M: It's still done, but each time on smaller and smaller animals. You can't do it experimentally on large animals like dogs, monkeys, cats. It's now unacceptable to carry out experiments on this kind of animals. It's restricted to large mice and guinea pigs at most.

I: What about using the computer to replace animal testing?

M: That's the great hope. We have to rely on experiments in a computing environment in order to substitute animal models.

I: Do you use the computer in your work?

M: We use the computers a lot because we have to simulate our models. Everything that we see as far as the dynamics of the project is concerned is carried out in a computing environment, so we can see how populations interact with each other.

I: In some sense, it's impossible to wipe out diseases; they usually morph into something else.

M: I think wiping out diseases is really very difficult. We should not aim for that. We have to better understand what is going on in these interactions between all these living beings, in particular, in the field of infectious diseases. I would not believe that we will eradicate all of them because there are so many new infections every week. The pathogens are evolving each time to be more infective, and we are evolving tools in order to be resistant. So this is a sort of arms race, and we have to keep maintaining the upper hand. That would be the aim.

I: You mentioned that vaccination is a sort of prevention, but what happens if the vaccine eventually becomes not so effective?

M: In some diseases, you have the vaccines, in particular those vaccines that apply living virus or bacteria, and sometimes it has a sort of reversion to the lethal state. We have vaccine side effects as well. It's not uncommon and it's a matter of much concern.

I: I remember that years ago people thought that tuberculosis has been wiped out, but recently they have second thoughts about it.

M: Yes, tuberculosis is still an infectious disease which kills the highest number of people in the world. It is the number one killer. We have about three million people dying every year from tuberculosis.

I: Is it mostly in the under-developed countries?

M: We have many problems in Africa and also in Russia, in confined populations like prison inmates and under-nourished children. It is still a global problem. Tuberculosis is still a challenge for us.

I: What is your advice to a student who wants to do research in your kind of area?

M: The advice is that you have to be prepared first of all, go after some formal courses on mathematics, statistics and computer science. Without these, it's very difficult even to interact with statisticians. The other way around, people from the mathematics background who wish to apply them should get some formal courses in biology. In order to apply at least the language to interact with other specialists, inter-disciplinary background is important. The second important point is to mix with people from other backgrounds. You have to work with other specialists. This is a team work. It is very difficult to survive alone.

I: Do you talk to the mathematicians back home?

M: Yes, we talk a lot. I have several mathematicians in my group. They are all learning biological science as well. We have to speak the same language.

I: It's quite difficult for a mathematician to learn biology, isn't it?

M: I would think it's more difficult for a biologist to learn the mathematics than the other way around. It's easier for a mathematician to learn biology than for a biologist to learn mathematics.

I: But still, it's not easy to convince a mathematician to consider your problems.

M: No, it's not that easy but, you know, if the initiative of the mathematician is in applications rather than in such diseases, it's not that difficult to convince him or her to study some biology.

I: You have been very successful in that.

M: I think I have had some success, no complaints about it.

Daniel

When I first developed it, one of the first applications I made was to transportation planning, predicting demand for transportation alternatives.

I: Were you commissioned to do the modeling or did you do it on your own?

M: I did that on my own, but I took advantage of the fact that there was research money available in the area because there was a new system under development. I was able to use the existence of that new system to get funding for a large project. Well, that original application in transportation has continued. Some of the most common uses of my methods continue to be in transportation. For example, in Paris and Hong Kong, it is used systematically, as I understand it, as an operational management tool to do things like real-time traffic management. I think it is also used for traveling planning. It's just a physical model — they are not using any deep theory. Like generic statistical tools, I don't have a single way of using it — people use it as they wish.

I: If you have patented those methods, it could have brought in some money.

M: Some people have told me that if I had patented it, I could have become wealthy. But I have a different philosophy. In everything that I do, I make it a point of giving it away freely. I'm a member of the open software philosophy, so all my software is openly available to everybody.

I: Are your methods part of the standard material in books?

M: Certainly in econometrics books, yes. They are standard procedures within most statistical packages.

I: In your research on behavior of choice, is there any finding that you consider to be counter-intuitive?

M: I would say that the most counter-intuitive thing that I have found (this is not my personal research but by a group of people in this area) is the finding that relatively high level economic decisions which seem to be a very complex cognitive activity involving a lot of learned activity and so forth seem to be tied to very primitive pathways in the brain — very direct connections to rewards in the brain. So you get the phenomenon that, for example, people will respond to play in lotteries or economic games in ways which seem to be very primitive, very fundamental in terms of their positions within the structure of the brain. That, to me, is very surprising. I would have thought that economic decisions would be very broad, high-level, dispersedly processed things but instead it is said that there are direct connections between economic decisions we make now and probably the earlier evolutionary development in humans of the pathways developed for reward or avoiding risk. It explains some of the strongest anomalies in economic behavior — the asymmetry between how people make judgments about gains and losses, the willingness of people to be altruistic and to trust other people. Some of these things are quite anomalous in terms of classical economics but when they study the pathways to reward in the brain, it corresponds exactly to economic behavior.

I: Could you tell us something about the Econometrics Laboratory of which you are the director?

M: My laboratory is primarily a service laboratory. My view of applied econometrics is that in the past, one of the limitations in economics has always been the difficulty of processing the large data sets collected. Economic data that is traditionally collected tends to be very large scale, like census data. Traditionally, economists are hindered in their ability to work with these; they had limited computers and limited skills. When I established this laboratory, my intention was really to provide a good background facility for economists to do large-scale empirical work and to use computationally intensive methods in econometric analysis. That's largely what we do. We have very large file servers, high powered computers and we service the large community of economists at Berkeley. Within the university, the laboratory is pretty open — certainly open to all members of the departments and students. When people not within the university need high powered computation, I try to accommodate them.

I: How often do you go back to your farm or ranch?

M: I have a little farm, about one hour's drive from Berkeley. I go there every weekend. When I'm there, I work very hard in my vineyard and garden. To me, this is refreshing. I grew up on a farm. Before I left for college, I worked very intensely on the farm. When I was young, I thought I would — I didn't plan to be a farmer because the work was too hard — I thought I would be doing something related to farming, like being a county agent. I always said that. Now I enjoy being back on the land.

I: When you are back at the farm, do you still think about your scientific work?

M: I do, yes. My own experience is that if you actually sit at a desk and try to prove theorems, sometimes you just go slower and slower. It's very hard to be completely linear in developing mathematical results. Sometimes I find that if I put a problem down, go out and work hard physically, then either in the following working or when I wake up the following morning, the solution is there. I don't think it (farm work) slows me down at all; it probably helps me scientifically.

I: Can you say you have found some insights into your scientific work while you were working on the farm?

M: Definitely; not because of the farm work I'm doing, but simply because, at least for me, when I'm trying to prove a theorem that is difficult or challenging, I often have to do it almost subconsciously. I have to work very hard to prepare my brain and then to make the final connections, I almost have to walk away from the problem and then the pieces come together. And the farm is a good place for that.

I: Are you optimistic about the economic behavior of human beings?

M: I'm certainly optimistic about our ability to study behavior. Behavioral science has made great advances — a lot of good tools are available and computers allow us to build better models. We have learned a great deal about experimental techniques and from doing experiments. Game theory is becoming an important empirical tool, it used to be primarily a theoretical tool. Empirical game theory is becoming very useful, and there are now very strong interactions between economics and biology (brain science). It raises the possibility of doing experiments in which we use biological treatment (hormone treatment) as an experimental device to study behavior. I think this is a marvelous opportunity to learn how the mind works.

I: Has anybody actually tried to make connections between certain type of economic behavior with certain activity in the brain?

M: Yes, very definitely. It's not my own research, but I have followed with terrific interest the work of Ernst Fehr at the University of Zurich. He's doing experiments in which people are administered particular hormones and then asked to play an economic game, some kind of ultimatum game, which involves trust. It's a game where the Nash solution — you don't trust anyone — is expected to be played but, in fact, people do not play the Nash solution. What he finds is that by changing the level of hormones in different treatments, you can drastically change the way people play this game. It is a striking demonstration of a direct link between buffer brain chemistry and human behavior — altruistic behavior, trust, social behavior.

I: Is psychology becoming physiologically related?

M: Well, I think it's becoming so, very strongly. I think that's a powerful scientific advance because it gives you so many more possibilities for good experiments.

I: In some sense, it's also a bit pessimistic that you cannot run away from certain aspects of the brain's malfunction.

M: It does suggest that there is a lot of chemistry involved in our tastes and in our behavior, and to some extent, the chemistry is heavily involved in the mind's function. On the other hand, if you look at it positively, what it suggests is that if we understand our own chemistry, there is a possibility that we can control it, hopefully for the good. This line of research raises some deep ethical issues. If one learns how to manipulate the brain and manipulate taste, it raises the ethical issue of whether one should do that or want to do that.

Keith

Magnetohydrodynamic Attraction

Keith Moffatt has, in a long and distinguished career, made important contributions to fluid mechanics in general and to magnetohydrodynamic turbulence in particular. His scientific achievements are matched by his organizational and administrative skills, which he devoted most recently to the Isaac Newton Institute for Mathematical Sciences at Cambridge.

Educated at Edinburgh University and Trinity College Cambridge, he first taught at Cambridge University and was Fellow of Trinity College from 1961. Except for a brief stint as Professor of Applied Mathematics at Bristol University (1977–1980), his career has been centered at Cambridge University, where he has been Professor (now Emeritus) of Mathematical Physics, Head of the Department of Applied Mathematics and Theoretical Physics (1983–1991), and Director of the Newton Institute (1996–2001).

He has been a visiting professor at the École Polytechnique, Palaisseau (1992–1999), Blaise Pascal Professor at the École Normale Supérieure, Paris (2001–2003), and Leverhulme Emeritus

Moffatt

Professor (2004–2005). He has served as Editor of the *Journal of Fluid Mechanics* and as President of the International Union of Theoretical and Applied Mechanics (IUTAM). For his scientific achievements, he was awarded the Smiths Prize, Panetti–Ferrari Prize and Gold Medal, Euromech Prize for Fluid Mechanics, Senior Whitehead Prize of the London Mathematical Society and Hughes Medal of the Royal Society. He also received the following honors: Fellow of the Royal Society, Fellow of the Royal Society of Edinburgh, Member of Academia Europeae, Fellow of the American Physical Society, and Officier des Palmes Académiques. He was elected Foreign Member of the Royal Netherlands Academy of Arts and Sciences, Académie des Sciences, Paris, and Accademia Nazionale dei Lincei, Rome.

He has published well over 100 research papers and a research monograph *Magnetic Field Generation in Electrically Conducting Fluids* (CUP 1978). Although retired from the Newton Institute, he continues to engage in research and to serve the scientific community. In particular, he is a founding member of the IMS Scientific Advisory Board (SAB), which has helped the Institute to find its direction during the crucial first five years and establish itself on the international scene. During the annual visit of the SAB, he was interviewed on 6 January 2006. The following is an edited version of the transcript of an interview brimming with reminiscences and good-humored chuckles and capturing the excitement of discovery in an important and very relevant field of scientific activity.

Imprints: You already had a first-class honors degree in mathematical sciences from Edinburgh when you went to Cambridge to do a BA. Were the first two years in Cambridge decisive in your choice of research area for your PhD?

Keith Moffatt: Yes, in fact my first year in Cambridge was decisive. In those days, it was still quite common for a graduate from a Scottish university to go to Oxford or Cambridge and take the BA. This was the tradition that I followed. I enjoyed fluid mechanics at Edinburgh University but I was also exposed to quantum mechanics, and I thought that my career would be in this subject — that was what attracted most graduate students in those days. It was related to nuclear research and everything else. In my first year in Cambridge, I attended more courses in quantum mechanics at graduate level, but realized in the course of the year that I didn't want to pursue research in that field. I yearned to go back to the fluid mechanics that I had enjoyed so much at Edinburgh. So after one year at Cambridge, I took that decision and went to see George Batchelor to talk about the possibilities in fluid mechanics.

I: That was in the Mathematics Department?

M: Yes, Batchelor was in the mathematics faculty, but he actually occupied an office in the old Cavendish Laboratory. He was quite close to physics as well. I remember very well my first meeting with him in his office absolutely crammed with books and papers. I had attended his course in fluid dynamics and I liked the subject, and turbulence was the natural area to go into. He was one of the world authorities at that time on the theoretical side. It was obvious that it was a very challenging subject, and it still is!

I: Was your interest in fluid turbulence largely due to the influence of George Batchelor?

M: Yes, he was the authority, and he had other very able people working with him in a strong research group — people like Ian Proudman and Philip Saffman; and G. I. Taylor still exerted a benign influence in the background. There was an atmosphere of great vitality in research in fluid mechanics. George put me on to a problem in turbulence. This was my second year in Cambridge when I started research, although I was still doing my BA.

I: Is it compulsory to do the BA in Cambridge?

M: Yes, I had to take the BA, although Batchelor took me on as a research student on the basis of my Edinburgh degree. He was an Australian and came from Melbourne University. He had a very open attitude (for Cambridge!). He regarded my degree from Edinburgh as quite adequate as a preliminary to research.

I: Did you do any experiments?

M: Not at that time. I was entirely on the theoretical side. I did some very simple experiments later in my career, but not on turbulence. One of the attractions in fluid mechanics is that you are concerned with phenomena that can be seen. You can easily visualize and that appeals to me. I like to do simple experiments. I like to watch, as we all do, the flow of water, for example, and the vortices that develop and the interactions of these vortices; it's fascinating. When an experiment can be easily done — a tabletop sort of experiment — then I will do it, often for demonstration purposes for students. It's interesting how often when you are preparing a demonstration for students, it raises more questions and leads to research problems.

I: As a child, were you already interested in observing physical phenomena?

M: I think most children are interested in what they see around them, they are curious about the behavior of mechanical things. It's one way to get children interested in science, trying to understand what we see around us. But, no, I think my real appreciation developed much later in life.

I: To be more specific, were you fascinated by the flow of water as a child?

M: Well, I always enjoyed water, I must say. Coming from Scotland, we were frequently on holiday either at the seaside or in the country where we have wonderful rivers and mountain streams. Yes, I would sit for hours watching the swirling flow.

I: Is magnetohydrodynamics mainly applied to astrophysics?

M: That is certainly one important field of application, but by no means the only one. Magnetohydrodynamics (MHD) has applications equally in geophysics, notably to the dynamo problem of generation of the Earth's magnetic field. Then there's the intensely practical problem of controlled thermonuclear fusion: the challenge is to contain a very hot ionized gas using a magnetic field. Many MHD problems arise in this context concerning existence, structure, and stability of magnetostatic equilibria. There has been huge activity in this area dating from the 1950s and 1960s. Then there's the whole area of liquid metal MHD, relevant for example to processes of flow control in the continuous casting of steel, and other metals and alloys. And in the developing of new materials, there's a process called crucible-free casting: you have to contain a sample of liquid metal in extremely pure form and you can do this by using magnetic levitation. There are many important practical applications of this kind.

I: You mention magnetic levitation. Some trains work on that principle.

M: It's a similar principle. But there you are levitating a solid structure. To levitate a fluid with its infinity of degrees of freedom, there are delicate problems of stability.

I: Are there any other practical problems?

M: Well, there's a host of stirring and mixing problems using magnetic fields. If you use an alternating magnetic field, for example a field rotating at high frequency, you can generate rotational flow in a container, and by carefully crafting the field, you can generate quite complex flow fields. If you are interested in mixing, this is a valuable technique, which is more sophisticated than using a spoon! Again, there has been a lot of work in this area since the 1960s.

I: Do magnetic fields have destructive effects, like radioactive particles?

M: No, they don't seem to have. On the contrary, the magnetic field of the earth protects us from very damaging radiation from outer space and it's very fortunate that we have a magnetic field serving as a protective blanket. I don't know what happens if the human body is subjected to an extremely strong magnetic field; it's not altogether known what the effects might be. It's better to avoid it. Experimental work in MHD can be dangerous not because of the strong magnetic fields that are used, but rather because you are dealing with very volatile substances. Even mercury is dangerous: it wasn't recognized in the '50s, but it's now well-known that the vapor from mercury is very poisonous, and so MHD laboratories using mercury have to be very carefully designed to meet health and safety regulations. Liquid sodium and potassium alloys are very high conductivity metals, which are used for experiments and are extremely dangerous, extremely inflammable. So you must avoid any possible leakage in an experiment.

I: Does every substance have a magnetic field?

M: Well, at the atomic level of microscopic fluctuations, yes. But in MHD, one is dealing only with fluids that are good conductors of electricity, either liquid metals or hot ionized gases.

I: How much progress has been achieved in fluid turbulence, at least in MHD?

M: Progress in turbulence at the fundamental level is extremely slow. You sometimes take one step forward and two backwards! This applies even to the most fundamental theoretical development in turbulence, the theory of Kolmogorov (1941) which essentially boils down to inspired dimensional analysis. Even Kolmogorov recognized a fundamental flaw in his theory, and he published a revision (his updated thoughts) in 1962, some 20 years later. At that stage, he himself undermined his own theory! One of the "firmest" foundations of turbulence from that point on became very shaky. This is typical of the history of the subject.

I: I think it was Feynman who said that turbulence was the major unsolved problem of classical physics.

M: I thought this went back to Einstein. You may be right, it may be Feynman. He was certainly concerned with turbulence in some of his writings. I think it is true to say that at the fundamental level, turbulence is still not fully understood. There are many approaches — mathematical, physical, engineering — and these are very different. You hope that there is some common ground at the center where real progress can be made. As regards MHD turbulence, the news is good; in fact, I think the greatest advance in understanding did come in magnetohydrodynamic turbulence and it came in the '60s. It came through what is now described as mean field electrodynamics where the turbulence is on small scales but you are concerned with evolution of the magnetic field on a much larger scale, so you have scale-separation, allowing you to average over the small scales and focus on what happens on the large scale. This works fairly well for MHD, and the application is very important both in astrophysics and geophysics.

I: Is it a statistical approach?

M: There is an averaging involved in it, so to that extent it is statistical, but it's a fairly rudimentary sort of statistics. You take care of nonlinear effects through this averaging but there is great subtlety in the process. The great leap forward was in this area. I was lucky to be involved through recognizing the relevance of a quantity called helicity in turbulence: this is the correlation between velocity and vorticity. It relates to distinguishing between right-handedness and left-handedness. The physicist would describe it as a measure of the breaking of chiral symmetry, and it is an extremely important concept in MHD turbulence. This realization developed in the late '60s and gained acceptance through the '70s; that was the great breakthrough. So in this area at least, we can look back on the last 50 years and say, "Yes, we have a big increase in understanding." But still now, when we look at pure turbulence, the undiluted problem with no magnetic effects, I don't think we have any such great increase of understanding.

I: You mentioned Kolmogorov's work on turbulence. Did Batchelor try to elucidate on his work?

M: Yes, this was his early work, just after the war, in 1946/1947 when he came to Cambridge and worked under G. I. Taylor, although he was from the beginning very independent. He unearthed Kolmogorov's papers from the bowels of the Cambridge library, studied them very closely, gave his own lucid interpretation, and gave them very wide publicity. It was through Batchelor's work that the theory became widely known in the West.

I: Do you think that in the next decade or so there will be conceptual breakthroughs in turbulence or do you think that computers will play an even greater role in understanding turbulence?

M: We must always remain hopeful! I think the answer is: "Both". Important progress in turbulence now does come, no one can deny it, from advances in high-power computer simulations. We need these, but computer simulations alone do not lead to real understanding. They have to be coupled with theoretical and experimental work. You really need a 3-fold interaction: computer simulation, theory and experiment. There will always be a place for careful theoretical analysis. Computer simulations often throw up new developments: for example, it was computer simulations that detected the prevalence of concentrated vortex filaments in turbulent flows. This immediately led to a new theoretical modeling and the search for an understanding of why these filaments are such a pervasive feature of turbulence. New theoretical insights then suggest new computer simulation experiments; and so on. We need both. As far as the next breakthrough is concerned, for the next 10 years (a good time-scale, I hope) there is one very big challenge and it relates to understanding the way that these concentrated vortices interact when they are non-parallel. In a fully 3-dimensional flow, they tangle with each other in a very complex manner. The big question is whether the associated solution of the Navier–Stokes equation remains smooth, regular for all time, or whether a singularity will typically develop within a finite time. This is an unsolved problem, which dates back to Leray in the 1930s. It is one of the seven millennium problems posed by the Clay Institute for which a prize of one million dollars is on offer. This calls for a mathematically rigorous solution, and that needs new theoretical ideas. The problem is that the computer can never demonstrate either a singularity in finite time or regularity for all time. At best, the computer can provide an indication of a trend, but then theory has to take over to establish that the trend is genuine and that it really does go to a singularity, or to regularity, or whatever. There is huge interest in this problem, which is central to turbulence because if it turns out that singularities of vorticity are a generic feature of incompressible flow, then there must be some means of resolving these singularities. My view is that compressibility must be taken into account on the very small scales at which such singularities occur. Compressibility effects are usually ignored and thrown out at an early stage in turbulence analysis.

I: Do the Navier–Stokes equations apply at the atomic level?

M: No. You do need to adopt a continuum approximation. Obviously that does break down when you get down to the level of fluctuations of density at the molecular level. Even so, the Navier–Stokes equations are still valid down to the level of microns, but not to the level of molecules or atoms.

I: One would have thought that at the continuum level it would be easy to solve the equations.

M: Yes, but it isn't. Within the continuum framework, the equations are nonlinear and dissipative. Also, the incompressible Navier–Stokes equations are non-local in character, because of the long-range influence of pressure. All these things conspire to make it very, very difficult.

I: Have the Navier–Stokes equations been modified?

M: Yes, they have been modified in a number of ways, depending on the context. For example, 2-dimensional Navier–Stokes is relatively easy; but in 3 dimensions, all hell breaks loose. You can creep towards 3 dimensions — 2½ dimensions, for example, where you take into account some 3-dimensional effects but not all. That's generally where progress is made.

I: I may be simple-minded, but going beyond 3 dimensions may help.

M: Oh, going beyond 3 dimensions to 4? That is possible. There are other examples in physics where you go to $4 - \varepsilon$ dimensions, where ε is formally a small parameter; then having done the calculation, you boldly set ε equal to 1, and you are back to 3. Attempts of this kind have been made in turbulence but so far have had very limited success.

I: What about fractional dimensions?

M: That has a bearing. There was hope in the '70s that new ideas from chaos theory would help to crack the problem of turbulence, but I think that was fairly short-lived. Certainly particle paths are chaotic in turbulent flow, and ideas from chaos theory are relevant to mixing, but they don't solve the dynamical problem of turbulence.

I: Historically, there seems to be a British tradition in applied mathematics (classical physics) that can be traced to Maxwell, Stokes, Reynolds, Taylor and Batchelor. Do you consider yourself to be a successor of this tradition, and how much of it is being continued?

M: Well, it would be pretentious to claim to be a successor of the tradition established by these illustrious names, but I am certainly a beneficiary! I was greatly influenced by Batchelor, and I had a close relationship with him until he died in 2000. I knew G. I. Taylor well also at Trinity College, till his death in 1975, and I had frequent opportunities to talk with him informally in the college. Taylor told me he had attended a lecture of Lord Kelvin in 1904; this is another name I would add to your list — a very famous name in classical fluid mechanics (I would also add Lord Rayleigh). Kelvin knew Stokes very well and they had an incredible correspondence that extended from 1846 until the death of Stokes in 1903. Through G. I. Taylor, I have this remote link with Kelvin and Stokes! Three years ago, we commemorated the centenary of the death of Stokes in Cambridge and I immersed myself in his papers in fluid mechanics and gave a lecture on this subject. I do have a strong feeling for the achievements of Stokes and his relationship with Kelvin. Of course, Maxwell comes into the picture, and I feel an affinity with him too. (Like Maxwell, I was born and educated in Edinburgh.) He was, of course, a very, very great figure in science, increasingly regarded as being in the same league as Newton and Einstein. What a tragedy that he died so young! That was right there in Cambridge, where he was first Head of the Cavendish Laboratory and Fellow of Trinity College. As regards Reynolds, he was a professor at Manchester, famous for his experimental observation of the transition to turbulence in a pipe and the fact that this apparently occurs at a critical value of a dimensionless parameter that later became known as the Reynolds number. So yes, I guess he's part of this great British tradition in fluid mechanics. I'm certainly happy to have been nurtured in this tradition.

I: What about its continuation?

M: I think it is still strong in Cambridge and the UK. We can't claim Kelvin at Cambridge because he spent his whole career as Professor of Natural Philosophy in Glasgow, although he had been a student at Cambridge and made frequent visits there, particularly to interact with Stokes. It is a peculiarly British tradition. The strength of fluid mechanics in the UK is a consequence of Stokes, Kelvin, Rayleigh and Taylor. Taylor didn't have that many students but his influence in the UK remains strong, particularly in my department (DAMTP) in Cambridge where a dominant theme is fluid mechanics and its many applications. I think this will continue but the nature of the investigations changes and, of course, the computer revolution plays an important part. You can't do research in fluid mechanics nowadays without being involved in computational work at the same time.

I: Are you still able to attract good students to do fluid mechanics?

M: Personally, no, because I reached the retirement age in Cambridge three years ago. I've had a very good research student from Poland these last three years (Michal Branicki) who has just completed his PhD, but it's not normal to take on new research students after retirement. It's possible but unusual. Most research students would wish to be with younger members of the faculty. There is a good continuing recruitment of research students into fluid mechanics in the department, certainly. Its applications are traditionally in the physical sciences and engineering, but the range of applications now embraces biological sciences, geophysics, and astrophysics as well; so it's very broad!

I: It used to be that the understanding of the term "applied mathematics" in the UK is different from that in the US. How much of this is it still so?

M: Well, I talked with Avner Friedman about it this morning. I agree with you that it used to be different, but the use of the term is now converging. Even within the UK, people would differ on what they mean by "applied mathematics". Even the distinction between pure and applied mathematics has been eroded, and quite rightly so. People don't like to use the term "pure mathematics" anymore, because some areas may be quite pure in one epoch and turn out to have important applications in the next. One of the functions of the Newton Institute is to surmount interdisciplinary barriers, particularly between pure and applied mathematics. It's one subject — mathematics and its diverse applications. For me, applied mathematics is mathematics applied to the physical and biological sciences. But some would extend the term to cover the social sciences also. Financial mathematics, for example, is that applied mathematics or isn't it? It's what you are practicing. The boundary between theoretical physics and pure mathematics has certainly been eroded. There is a very strong interplay between the two fields now and they are mutually beneficial.

I: The physicists do not seem to be very happy about that.

M: But there are some brilliant exponents like Michael Atiyah who started in pure mathematics but who gradually embraced theoretical physics. Maybe that's what the physicists are not too happy about, but it's a fact of life. Theoretical physics is a close partner of applied mathematics. My department back home is the "Department of Applied Mathematics and Theoretical Physics", reflecting that these two disciplines are separate but related; the boundary is flexible!

I: I think that in the US applied mathematics is more about applications outside the physical sciences.

M: Perhaps you have in mind applications to economics and the like. That is possible, but Avner would dispute this, I think. Perhaps the interpretation of the term has changed in the United States. It is difficult, I agree with you. Someone in the States working in fluid mechanics would be more likely to be attached to a Department of Engineering than a Department of Mathematics. It is regarded as being more within the ambit of engineering. In Britain, we succeeded in keeping this kind of applied mathematics — fluid and solid mechanics — within faculties of mathematics. In this way, top mathematics students can be more easily attracted to the subject.

I: Did you have many graduate students during your career?

M: I did have a steady stream of graduate students and this is one very rewarding aspect of an academic career. I have been very fortunate to have had many good graduate students in my time. They are all good but some of them are absolute stars — people like Juri Toomre, Andrew Soward and Michael Proctor, who have done extremely well in their subsequent research careers. It is rewarding to work with graduate students, when mentoring develops progressively into collaboration. That is why it is so satisfying and it keeps one young. The new students coming in, of course, get younger and younger! The fresh excitement with every new graduate student is something very rewarding.

I: You succeeded Sir Michael Atiyah as the Director of the Isaac Newton Institute for Mathematical Sciences in Cambridge. What is your most satisfying accomplishment during those five years as Director?

M: I inherited from Sir Michael a very exciting development — the Newton Institute was five years old when I took over — the appointment as Director is for five years. The primary role of the Director is to maintain a high level of scientific visitor research programs and I think I did this during my five years, and this still continues. But as well as maintaining the high scientific level, the Director has the responsibility to maintain the financial health and viability of the Institute, and that is not easy. I was very much concerned with the financial health for the longer-term future of the Institute. I was in constant debate, and indeed argument, with our National Research Council (EPSRC) to maintain an adequate income level, and equally with Cambridge University and its Colleges for internal support. I was also constantly seeking to raise money from external private sources, and I was reasonably successful in this. During my time, we won one of the Queen's Anniversary Prizes for innovation at the Newton Institute, on behalf of Cambridge University. And to celebrate the millennium year 2000, we produced a series of 12 posters to demonstrate the enormously wide scope of applications of mathematics. They were reprinted by World Scientific in Singapore and have been widely distributed in Southeast Asia as well as in Europe. We have reproduced them in this little booklet. It gave a fair spectrum of the applications of mathematics. That was quite exciting and involved a lot of work during 1999 and 2000.

I: Were they distributed to the schools?

M: Yes. They were designed for display in the trains of the London Underground with the general title *Maths Goes Underground*. Each month a new set of posters appeared in the trains. After that they proved popular and there was great demand from schools. So we reprinted and distributed to all schools and universities in the country. They were all over the place.

I: Our own Institute for Mathematical Sciences is modeled partly after the Isaac Newton Institute. What do you think are the similarities and differences between these two institutes?

M: I was first aware that IMS was to be modeled to some extent on the Newton Institute when your Deputy Prime Minister, Dr Tony Tan, visited the Newton Institute in 1998. He came with Louis Chen, looked carefully at what we were doing, and we had a long discussion. This is how I became involved in IMS. There are similarities — the idea of holding programs and bringing in visitors from overseas, this is at the heart of the business of any visitor research institute: short-term programs of up to six months duration, with as many distinguished visitors as you can attract to come and engage in research, and interact with the local community. That is very much the spirit of the Newton Institute also. As regards the differences, the Newton Institute has a wider catchment area — the whole of Europe is at its doorstep. Many of the participants and many of the young postdocs and graduate students come from Europe. It's now very easy for Europeans to fly into Stansted Airport near Cambridge from anywhere in Europe. So there is a very large community there. It's not only Europe, of course. We have many visitors from the United States and from all over the world, but primarily you look to your local community. Of course, Singapore has a strong local community but it is relatively small. You have a wider Asian community. I think that IMS must regard itself as a beacon for that community, and extending to Australia. If you look at the globe, you can see that Singapore can be an attractor in a certain area. You have currently tremendously strong growth from China in particular.

I: But the local scientific level is lower…

M: Well, perhaps, but the ambition of IMS must be to raise that level to reach equality with the institutes in Europe and the United States. I think it's doing very well in that respect. The level of the programs here has been high. I think it is more difficult to maintain that level of activity here given the geographical isolation and the fact that you've got a smaller community in Singapore itself. The first five years is always easy. The second five years and the third five years — to maintain sustainability — will be more difficult. There is a danger of running out of steam, you know. There is quite a problem there of just keeping it going at the required level of intensity.

I: Has this to do with the "critical mass" needed?

M: Yes, it is a question of critical mass. There is also the question of diversification. "Mathematics and its applications" has to be interpreted in a broad sense. Any area of mathematics and its applications may be a potential area for a good concentrated program. The programs here tend to be of a shorter duration. It's difficult to get people to come and stay for more than a month or two. They'll come for a couple of weeks or one month for workshops, but to stay for a longer extended period is quite difficult. To maintain a research activity for up to six months is not easy. Two months seems to be workable and a good compromise. If IMS runs four or five programs in a year, each of two months' duration, I think that's excellent and can work very well.

I: Do you think we should be focused on certain topics rather than spread out over a large number of areas?

M: Well, despite what I just said, I do think it's good to focus here on topics that are most relevant to Singapore — local problems. It's interesting that environmental problems are emerging as one of the key areas. Environmental fluid mechanics is important in relation to problems of pollution, and problems relating to natural hazards. That tsunami was so close to Singapore that it must have been a matter of great concern here. Phenomena relating to extreme weather conditions are of ever-increasing concern, and these fall well within the scope of mathematical investigation. And then there is the whole vast field of biomedical science — another area in which Singapore can make great contributions. This is an area also where mathematics can play a vital underpinning role.

Stanley

Mathematician with an Edge

Stanley Osher is an extraordinary mathematician who has made fundamental contributions to applied mathematics, computational science and scientific computing and who has cofounded three companies based, in part, on his research. He has applied his pioneering work on level set methods and other numerical methods for partial differential equations to the field of image processing and, in particular, to video image enhancing and movie animation. He has been featured prominently in the scientific and international media such as *Science News*, *Die Zeit* and *Los Angeles Times*. He is perhaps the most highly cited researcher in the field of scientific computing.

He received the NASA Public Service Group Achievement Award, Japan Society of Mechanical Engineers Computational Mechanics Award and the SIAM Pioneer Prize. He was an invited speaker at the International Congress of Mathematicians.

He is currently Director of Special Projects at the Institute for Pure and Applied Mathematics (IPAM) at the University of California at Los Angeles and Director of Applied Mathematics.

Osher

Stanley Osher was interviewed on 17 December 2003 when he was an invited guest at the Institute's program on imaging science and information processing and gave a public lecture. The following is based on an edited transcript of an interview in which he talked about the fun and fascination of applied mathematics and his total dedication to research and applications.

Imprints: In which area did you do your PhD?

Stanley Osher: I did my PhD in an esoteric area in functional analysis, which is in pure mathematics. I left it immediately and switched to numerical analysis after my thesis. I was lucky enough to talk to some people, including Peter Lax, who suggested the numerical stuff.

I: Did you find your real inclinations in applied mathematics?

O: Yeah, but it did not happen until after I got my degree. I liked everything and I specialized more after my PhD.

I: Did you use functional analysis later on in your work?

O: Yes. The first thing I did in numerical analysis was an application of Toeplitz matrices. It used functional analysis and was short and elegant.

I: You switched to applied mathematics and then eventually to something very practical like applying to movie animation.

O: This just happened — that's the way research leads you. You cannot predict these things. Together with colleagues and students, I was using the level set method, which is a way of determining how surfaces such as bubbles move in three dimensions, how they merge and so on. You can simulate the flow of bubbles, planes and things like that. It happened that people in the movie industry got interested in this stuff.

I: Are these pure mathematics problems?

O: It's a way of representing surfaces and has connections with differential geometry. People prove theorems about these things. They have applications in many areas including fluid dynamics and quantum mechanics. They arise in the movie industry because you want to see how things merge and split like in explosions or rising bubbles. You know, our Governor, Schwarzenegger, used in his latest movie a lot of these methods done by my former student Ron Fedkiw.

I: How did you get into the movie animation business?

O: We had a week of movie industry people coming into UCLA (University of California at Los Angeles) giving lectures about what they did and, in fact, imaging science was highlighted by the American Mathematical Society one year. There was a week in that stuff. We invited people from the local movie industry and they were interested in what we were doing. They wound up arguing with me.

The water in the *Titanic*, which won many Academy Awards, was very bad. It was old-fashioned stuff. Their people came to talk about that and so we decided we could do better than that. In recent movies, the water is much more realistic. The first movie that actually used sophisticated water was *Antz*. Now level sets are used by movies like *Shrek*, *Terminator* and many blockbusters. My former student Ron Fedkiw is doing this movie animation stuff very well. The stuff that my colleagues and I do has applications in many other places besides the movie industry.

I: I believe that another of your dramatic achievement is the use of mathematics in catching criminals. How did this come about?

O: Well, I was living in Los Angeles when the city went up in smoke. There was a big riot in Los Angeles after this guy (Rodney King) was beaten up by the police. The riot resulted in people being arrested for looting and beating up passers-by. There was a video recording of the bad guys beating up truck driver Denny and it showed a speck on the arm of a man throwing a brick at Denny. It turned out that I had a friend who knew the District Attorney or somebody, and I was then doing video image enhancement with my colleague L. Rudin. We were able to resolve the speck into a rose tattoo and it was a great application of what we were doing. After the Denny case trial (the tattoo led to the conviction of the suspect) we had a lot of media publicity and our company specialized in the area of image enhancement. Eventually, I sold my share of the company to Rudin. He has a package on video image enhancement which is used by the police around the world, and he's quite successful. I have left this business. It was quite fun and related to mathematics. Image processing is the real world and the graphics that is manufactured is the fake world. You want to find out what the image really is.

I: Do you actually go and seek out those problems or do people come looking for you?

O: It's hard to say. Sometimes it's serendipity. Things just happen. I was lucky. In all my years of doing science, I managed to work with the right people who knew what the problems were. For example, I knew nothing about image processing at all after my PhD. Then this guy Rudin came over to me and asked me about some work I had done in fluid dynamics on supersonic flow and shock waves. I asked him what he wanted to know and I got fired up. He was a computer scientist and he realized that shock waves had something to do with imaging. It was a fantastic observation and our collaboration worked out well.

I: The scope of movie animation has just opened up, hasn't it?

O: Yeah. But I'm not sure this is the best field of application for somebody to work in because the market is small except for video games which is a big business. But video games require real time imaging. What we do is not real time; it's too slow. That might change.

I: It could be just a matter of computing power.

O: Yeah, yeah, and also hardwiring in level set methods, which may come in time.

I: Pattern recognition used to be sort of very big.

O: I'm not sure of the definition of pattern recognition, but image processing is related to it. The key idea in everything we do is about "edges", which characterize images. Edges, and now, textures. If you look at a table, for example, the flat part is not very interesting but the boundary of it is. It's the discontinuity. If I look at you, I can see you because of the outline. The outline is very important and the mathematics that was used in other areas of science like fluid dynamics specialize in things like edges and boundaries. And now images.

I: Does it mean that if you have a vague or blurred object, you can always refine it?

O: Yeah, but it's usually difficult to do so when you have edges because near the edges you will get spurious oscillations. The techniques we have developed, which came originally from fluid dynamics, are now useful for removing those unreal artifacts.

I: How do you know that what you get is the actual object?

O: That's a good question. But you can get enough science behind it and you can prove theorems, algorithms converge and stuff like that.

I: You look at the picture and it's so blurred. And then, you do this and you get that. There's a lot of faith in that.

O: I won't disagree with that completely. But when I drop something, I expect it to fall and not go up. The probability is not zero that it might go up. But with a high degree of certainty, you can say that this is a realistic picture.

I: What about problems in voice recognition?

O: It's a different kind of mathematics. I'm not an expert in it. The techniques we use in image processing involving differential equations have never been used for sound. But I understand there's some very new stuff along these lines. It's only beginning and just developing. The Institute for Pure and Applied Mathematics (IPAM) at UCLA is running a program on sound and how the ear works and the related mathematics.

I: The imaging business is also very important in astronomy, isn't it? They take pictures which are so faint.

O: Very much so. Astronomers have done very good work in this area. They had to over the years. The early work in this stuff in astronomy was very fascinating. They did things which are precursors of what is going on now.

I: You mean the astronomers actually did something mathematical?

O: Yes. That's very often the case. The good thing about being an applied mathematician is that when you work in different areas you find very brilliant people in other areas of science develop mathematical algorithms without realizing what they are doing and which can be generalized to other areas. So, it's a question of language.

I: Are more astronomers talking to mathematicians now?

O: Actually we'll be running a program on computational astronomy in IPAM a year from now. Astronomy is just one example. Throughout science, mathematics is playing more and more of the key role. In the Institute in which I am involved, its mission is to do interdisciplinary work. People from different areas of science have problems which they think are mathematical. And our goal is to make mathematics out of it. Now they believe that we can do something, mainly because of the computer.

I: That's interesting. So now mathematics is also contributing towards understanding the origin of the universe. This is something not many people are aware of.

O: Well, I'm not an expert in the field, but yes, absolutely. I think that the world is governed by differential equations.

I: Your work involves a lot of algorithms. Do you invent the algorithms?

O: That's what we do. That's the fun part of it. I was interviewed by the *LA Times* (so was Tony Chan for the same article) and I said that I wrote the algorithms that make the computer sing. I am the Barry Manilow of mathematics.

I: Talking about algorithms, some people consider algorithms to be inventions.

O: Yeah, they are. You can actually patent them now. It used to be that a patent has to contain a device with wires and everything. But I understand the US Patent Office is now more liberal. I'm no lawyer, but I know that as the years go by, the Patent Office seems more and more interested in giving legal protection to algorithms.

I: You could be rich. Hollywood would be paying you millions.

O: People work for salaries. There is money, ego and fun. It's a very nonlinear function. I don't know which is most important. Fun is very important. It's a very good life. I would recommend people going into this stuff now. If you have the talent for it, it's the best life.

I: It's something very different, something, how do you say, non-academic?

O: In some sense, yes. You learn things, you read stuff and you learn new ideas, and you are fired up. Sometimes you deliver something different from what you have found. You have a vague idea that something interesting is going to come up. You wander around and something happens. Then you get very excited. It's like opening a door and you don't know what good things are behind it. You're not sure where it's going to end and what the level of success it's going to be. It's very exciting. Everyday I can't wait to go to work. People often asked me, "What kind of life is this that work is so important?" People go on vacation. My work is vacation.

I: Do you have many PhD students?

O: Many, disproportionately Chinese. We have many good Chinese students at UCLA.

I: Have you used your methods for something more serious like weather forecasting and earthquake prediction?

O: Yeah. The differential equations and some stuff I used to do in fluid dynamics and done by many other very good people are absolutely useful in weather prediction. I still do work on explosives and multi-phase flows and ray tracing. Physical phenomena apart from imaging is very much a part of my research with my colleagues and students.

I: What about the theory of turbulence?

O: Ah, turbulence is too dangerous. If you touch turbulence, you get burnt. One of my mentors once said he had great respect for people in turbulence, which is far more than they have for each other. Turbulence is too controversial. Turbulence has a probabilistic aspect to it, it's statistical and that's not my thing.

I: I'd like to ask you a philosophical question. How does your work affect your view of life?

O: In terms of how research affects my philosophy? The basic idea is to try to make order out of this life that we live. Everyday you encounter things and it's a messy world. The goal is to take this mess that we see and somehow "mathematize" it and make a prediction. In that sense, research has certainly affected my philosophy. I try to figure out what is going on. The most complicated thing is how our human nature operates. It will be fun to understand that. Many people I know at UCLA and elsewhere are using medical imaging to understand how the topology and shape of the brain affects its function. They use the mathematics that I am involved in. The greatest mystery of all is human behavior and maybe it can be explained by level sets.

I: Would you say that your present interest and activity is, in some way, directed by your own personal philosophy towards finding order?

O: Yeah, absolutely. I came from being a poor boy in Brooklyn. I wanted some order in my life, to become middle-class and to have a life that I enjoy. Then I stumbled onto this thing, and wow, that's very good for creating order. I entered graduate school in New York University in 1962 when it was a fantastic place for applied mathematics, maybe one of the best ever in the world. All the top people from Goettingen wound up in New York and it was so exciting.

I: Was that the Courant Institute?

O: Yes, the Courant Institute. In 1962, when I entered, it was incredible. The people who were there and the atmosphere. You felt that you were doing something important. It had a very great influence on me and many other people.

I: Was Courant still there?

O: He was still around, old but still functioning. He added many people all of whom I thought of being old, something like 20 years younger than I am now. They were great people and it was a fantastic time. There were many people like me who were ethnic New Yorkers and who view becoming a mathematician as a way of becoming middle-class American citizens.

I: What attracted you to UCLA?

O: In truth, there was a guy whom I was working with: Andy Majda. He is an excellent applied mathematician. He was there and they were building an applied mathematics group. Also I liked California. When I was in New York, I was always dreaming about the Beach Boys. Sunshine and California together with math was great — everything that I wanted. Over the years we had very nice people. The atmosphere is extremely good in UCLA. People who visited us commented on how well we get along with each other, which is unusual in academia. It has worked out quite well.

Doug

Computer Vision, Digital Magic

Doug Roble is world-renowned for his important contributions to computer vision and computer graphics and for pioneering applications to movie special effects and animation.

Roble did a joint degree program in engineering and computer science at the University of Colorado and went on to Ohio State University (OSU), where he did his PhD in computer science (on computer vision). He was an assistant faculty at OSU for a year before joining Digital Domain in Venice, California as a software engineer in 1992. Expanding on his PhD work, he developed the 3D tracking software TRACK for camera position calculation and scene reconstruction. This helped artists determine where to best fit graphics into images that have been filmed. For this software, Roble received a Technical Achievement Academy Award (Academy Certificate) from the Academy of Motion Picture Arts and Sciences in 1998. His subsequent work in the development of fluid simulation system earned him, together with Nafees Bin Zafar and Ryo Sakaguchi, a Scientific and Engineering Award (Academy Plaque) in 2007. This work allowed graphic artists to create large scale surging water effects for the

Roble

movies *The Lord of the Rings: The Fellowship of the Ring*, *The Day After Tomorrow* and *Pirates of the Carribean: At World's End.*

He has been the Creative Director of Software at Digital Domain since 1993. He is the Chief Editor of the *Journal of Graphics Tools* and is on several panels and committees, including its Advisory Board, of SIGGRAPH (Special Interest Group in Graphics), the most prestigious computer graphics conference. He has given invited lectures and keynote addresses at many major conferences, most recently at the Annual Meeting of the American Association for the Advancement of Science in 2007. He received the Distinguished Alumnus Award from Ohio State University in 2002. He is a voting member of the Academy of Motion Picture Arts and Sciences, Visual Effects Branch.

Roble was one of four invited speakers at the *Symposium on Mathematics and Science in Digital Media, Technology and Entertainment* held at the Raffles City Convention Center on 1 July 2007 that was organized by the Institute jointly with the Department of Mathematics, NUS. The symposium was supported by the Media Authority of Singapore to introduce and publicize the new field of interactive digital media to the general public. Roble was interviewed during the symposium. The following is an edited and unvetted version of the transcript of the interview in which he spoke with passion and animated enthusiasm about his early work on computer vision and its subsequent breathtaking impact on digital media and the entertainment and movie industry.

Imprints: Your BS was in electrical engineering and computer science way back in 1984. Was it some kind of joint program or major?

Doug Roble: It was a joint program actually because there was no full computer science degree in the University of Colorado at that time. The only way to get into computer science, which I knew I wanted to pursue, was to do an electrical engineering and computer science degree. Also I wasn't sure what I wanted to pursue — I knew that electrical engineering was interesting as well. It turned out to be a good thing because an electrical engineering degree offered much more math than a typical computer science degree, especially at that time. Computer science at that time sort of required you to have linear algebra, maybe a little bit of Boolean math, but electrical engineering gave me a good foundation in calculus, multivariate calculus and signal processing. I've come to use a lot of those basics much more than I thought I would. So it was a joint degree, a sort of double major.

I: Was a double major common in those days?

R: Kind of. Computer science was an emerging field. If you wanted to get a computer science degree, that was the way to do it. There was no pure computer science degree at that time, as I recall it. I think it was only a couple of years later that the University of Colorado had one.

I: What attracted you to Ohio State University subsequently to do your graduate studies in computer science?

R: Remember that, way back in 1984, Ohio State University had, and still has, a very strong graphics program. Back in 1984, they were associated with a company called Cranston/Csuri Productions, Inc which was a pioneer in computer graphics. They did a lot of the first commercials using computer graphics. In fact, they had done this famous commercial with a very shiny robot woman talking about the beauty of canned food. It was associated with Ohio State University. It was one of the pre-eminent computer graphics school at that time. Stanford hadn't even started its computer graphics program at that time. Ohio State was doing it, so I wanted to go there.

I: Did it ever occur to you to take up engineering instead?

R: Well, no. After my bachelor's degree, I used that time to figure out what I wanted to do. I knew it was computer graphics. I was fascinated with what I saw was happening with the movies. Remember it was 1984. *Star Wars* and *The Empire Strikes Back* have already come out. *The Return of the Jedi* had just been finished. I wanted to do that. *ET* and all those great films that were using traditional effects could have used computer graphics as well. This is what I wanted to be doing.

I: After your PhD, you were in academia for less than one year in OSU. Was it a calculated plunge to go into the digital industry at a time when digital media was at its infancy?

R: Absolutely. I knew I didn't want to be in academia. I wanted to work in films. It was a good opportunity because it allowed me to jump into the beginning of the bit of the domain. The company [Digital Media] had just been formed. It opened its door in 1993 and I was its 31st employee hired. So I was right there at the beginning, and it was a bite of a bullet, and I was a bit scared that the company might not last very long. But it did work out fine.

I: It must be quite fun to start at the beginning.

R: It was. It was amazing. It was crazy.

I: How recent is this discipline of IDM [interactive digital media]? How do you define it?

R: When you put the word "interactive" in front of "digital media", it becomes a whole different thing. Interactive digital media tends to mean games graphics, maybe even visualization. I'm in digital media, not so much interactive digital media, which really started in the nineties with video games and things, when *Doom* and the first 3D games came out. Now movies and games are coming closer and closer together. It's all so crazy. I tend to define interactive digital media as that where in response to the user's input, something on the screen changes. I work in films where it doesn't matter what the user is doing. The user can leave the room and it still gets projected on the screen. The problems we are trying to solve are vastly different. Most of the things we render take ages and ages to render, from an hour to 24 hours. With video games, you work with 60 frames a second or you're in trouble.

I: Some time ago, there was some kind of movies where the audience actually participates in choosing what is going to follow.

R: There was this little teeny experiment where you get to choose between (1) and (2) endings. It was just an experiment. Maybe it will change sometimes but you don't want the majority to win. I think much more likely you will have DVDs where one person gets to choose the plot rather than the majority who's watching a film. It's democratic but that doesn't make sense.

I: Could you tell us something about your most exciting research work? Is your PhD research related to your later research work in industry?

R: Indeed, my PhD work was in trying to use computer vision to help computer graphics. I took the basics of that and re-did it for Digital Domain when I first started the program TRACK which is a computer vision toolkit that basically allows artists to look at an image in some film and extract as much information as possible — where the camera was, what the scene looks like, all the 3-dimensional information you can possibly get from a photograph. This has been something that I've been working on for the 13 years I've been there. I continuously backtrack to add new features to it. It got me the Academy Award. That was the best thing that could have happened and was probably my most exciting research work although the third generation stuff that I am doing is very, very cool. It's such a visceral feeling when you see things that are flowing like water and look like water. It's very fun. Right now, I'm looking at all sorts of stuff. One of the things that interest me right now is hair. That's probably my most current research work along with other people at Digital Domain. By the way, nothing happens by yourself. It's all part of a team. There's a group of people working on hair and it's fascinating.

I: What do you mean by working on hair?

R: Well, hair is a big deal. There are three aspects of hair. There's modeling hair, styling hair — putting hair on some head or body so that it looks like a human head. Once it's in that position, you want to animate it, simulate it so that when wind is blowing or when somebody runs his hand through the hair, the hair moves correctly. And third, render hair. How do you render hair with a hundred thousand very thin little strands that are semi-transparent and light bounces off it in a very interesting way? How do you render it so that it looks correct? By the way, everybody has different hair than you do. It's all very different.

I: It sounds extremely computationally difficult.

R: Exactly. It's a huge computational problem — a hundred thousand strands of hair, each one continuously curving — simulating them in a discrete fashion is very difficult. So you have to make simplifications and adjustments that are good enough so that the audience is fooled.

I: Do you actually apply the laws of fluid mechanics to the motion of the hair? It sounds incredible.

R: Absolutely. It's hard — it goes back to things like Cossart curves, ringed together in a chain to represent continuous curving segments. This was invented back in the 1930s to deal with bending objects. Hair is wonderful, it doesn't stretch though. You got to make sure that the hair does not stretch and building that into the math is important.

I: Did you have to invent some new concepts or techniques to resolve some of these issues?

R: We're working on it. A group in France has laid down the foundations. We're trying to take some of their ideas and modify them so that we can use them. We're not finished yet. Maybe ask me again in a year. Indeed, we invented some cool stuff.

I: Has anybody written some kind of foundational textbook on such things?

R: Not so much. Actually there is a very good researcher in Switzerland, named Nadia Magnenat-Thalmann. She's been working on clothing and hair for her entire life, and has been doing some very good research answering some very good questions. She has a paper in a book on hair and clothing simulation. It's a computer graphics book with a lot of math in it.

I: Were there any IDM problems that contributed significantly to the development of any area in mathematics or computer science?

R: Ah, things that feed back into computer science and mathematics. Absolutely. From me, not so much. I haven't really had a lot of impact outside... well, some of the stuff I have done, things like fluid simulation. We started looking at fluid simulation — basically, the computer graphics community, not just me — in terms of how to create water that looks realistic, which was a completely different approach from what computational fluid dynamics people were doing. They wanted to model water or fluid in a very precise way. Towards that end, they had to simplify the problem constraints because you can't model water realistically if you've got a very complex domain. We looked at it from an entirely different angle. We didn't care that much if it was totally realistic, but we wanted to put it in a very, very complex domain indeed. We have arbitrary boundaries, moving boundaries and all those stuff, and we wanted water to look real. We stood on the shoulders of a whole bunch of computational fluid dynamics rather than feeding back into it because we approach the problem in a totally different way. So we have attracted some of the attention of pure fluid mechanics people. And they said, "Oh, you're doing it that way. That's very interesting." So, Stanley Osher, Tony Chan invented level sets. I don't even know if they realized how important it was going to be. They took it and applied it with Ron Fedkew to fluid, and this is a brand new field. It's a new way of doing it and the computer graphics media have adopted level sets to do all sorts of amazing things, and that has gone back into the mathematics. I think that's one example.

I: What about the classical Navier–Stokes equations? Any contribution to it?

R: We use a subset of the Navier–Stokes equations. The ones for inviscid fluid are pretty puzzling. For the ones we do use, we are trying to push solving them faster and faster. Also, we have pushed ahead trying to capture the details. Whenever you are solving the Navier–Stokes equations numerically, there is a lot of filtering going on. You always lose details. All this stuff get lost in the mathematics of fluid. We recently (when I say "we", I mean the computer graphics teams — people in Berkeley and Stanford are really leading the way) are coming up with ideas of putting back the detail into the fluid simulation so that the detail isn't lost, or if we do lose it, we put it back in a possible way so that it looks good. The goal is always to try to render water that looks exactly like water. Water is very non-viscous, and that kind of fluid simulation is very hard to do. We're getting closer; we can do milk. Milk is easier, it's viscous and doesn't have all the sharp edges that water has.

I: Are creative computer programming skills necessary for a successful career in IDM? Can such skills be taught to any beginning mathematics graduate student?

R: Sure. Thinking in terms of math is very similar to thinking in terms of computer programming. When we hire new people, no matter where they're coming from, we expect that they know how to write code. Teaching programming to mathematics students is such an important part — using the computer nowadays, no matter what. That's just Mathlab or Mathematica. But having to write your own piece of software, to implement something you have done, at least in applied math — I don't see how you can get by without it. It gives you that ability to say, "Oh, I wonder if I can write this as a code and see it on a grand scale."

I: There are people who seem never to be able to write computer programs.

R: And they can do proof? What? No. Computer programming is easy, come on.

I: Certain computer programming is not as straightforward as proof and can be quite tricky. Don't you think so?

R: I don't believe that. When it was first invented, it was tricky. You have to be very careful. When I was first doing computer programming, I did it on ancient machines. You have to do it in assembly language and it was all very arcane. Now there are so many tools available in computer programming. Modern C++ languages or Java or any kind of programming language gives you so many advantages that once you learn the basics of looking at a problem — you learn iterative, looping statements, recursion, you learn how variables work — all of a sudden, you are writing code. All these things are mathematical concepts. It's not hard. You do use them when you are doing proofs.

I: The older generation of mathematicians somehow or other seems to loathe computer programs. There's a perception that computer programming is a young man's game.

R: You can do it, you just don't want to. What younger generation writing codes. I won't buy it. Certainly, young men are doing it. But you can do it. No excuses. You're being lazy [*laughs*]. If you are good at math, you can do computer programming.

I: Reconstructing a 3-dimensional object from a 2-dimensional image like a photograph seems amazing, if not unbelievable. Is it theoretically possible for known methods of reconstruction to fail in at least some contrived cases?

R: Oh, absolutely. If you just have a single image of a 3-dimensional scene (you just take one photograph), it's impossible to figure out what's going on. There's no way, without any extra information, to know the 3-dimensional nature of that scene. Even if you have multiple photographs of some scene, where you can do triangulation and the various computer vision techniques to figure out what's in that scene, there are still things like scale invariance. There's no way of telling whether the photograph of the fire truck you are taking a photograph of was a real fire truck or a toy fire truck. You can't tell the difference without some measurements you actually took at the scene.

I: How many photographs do you need to reconstruct a solid object?

R: You usually get away with two. Three helps. Of course, you can always reconstruct the object that you see, you have to have coarse refinement between the two. If the camera didn't move very much, then there are limits to the accuracy because the pixels are a discrete measurement of the world and there is a built-in error. Computer vision is all about managing the error. So if you identify a feature within an accuracy of a pixel or two, and then you move the camera only a little wee bit, then the error involved in that feature identification overwhelms the mathematical induction that you can do. So the result that you get is not so good. But if you have a couple of, or multiple, photographs with decent baseline, then you can do amazing reconstruction nowadays. But you have to be able to identify a correspondence between a feature on one photograph and the feature on the other photograph. So if I take a picture of a chair, and then I move the camera and take another picture of the chair, there are parts of the chair I can't see that I could see in the first. So those pixels are fine and I have to infer the details and kind of make them up.

I: This must have been applied in astronomy.

R: Oh, of course. That's exactly how they determine how far the stars are. You use a telescope and wait a couple of days for the earth to have moved, and then you get a very long baseline and you can use triangulation.

I: The first animation in films was based on a frame-by-frame representation. How does the current animation in films differ qualitatively, and not just quantitatively, from those earlier ones?

R: Well, first of all, your question ignores one of the aspects of cinematography of the early animation. It wasn't frame by frame representation; it was frame by every other frame representation. If you go back to old Disney cartoons, hand-drawn Disney cartoons, because it was such work to animate every frame, to draw picture for every frame, Disney said, "Okay, we don't have to do that; we will only animate every other frame." Animation on two. If the motion is very rapid, sometimes you have to draw animation on one, where you actually draw a separate image for each frame. If you look at the old Walt Disney films that were poorly animated, you can see the difference. If you step through it on a DVD, you will see the fact that the images were held for two frames and then they move to the next frame. So right after that, qualitatively and quantitatively, we now animate every frame. Because the computer is doing it, it's pretty much as easy to do it on one as it is on two. There's no advantage doing on two. In fact, it looks smoother. But other than that, the technology of computer graphics has to do a lot of other things that you could not do on a hand-drawn animated way like motion whirl. Every film that is done nowadays renders not only the image but the image as it is moving, and so you get motion whirl. Without that, it looks very harsh and rigid. You pick it up and say, "Oh, it's fake. That looks like a computer-generated thing." With motion whirl, that makes it look more real. Unfortunately, motion whirl is very expensive. It's a sampling problem. It's a big signal-processing problem where now not only do you have the image rendered, but now you have to move it through time in order to get that broiled. And time is very continuous. Whenever you hear the word "continuous", you go, "Ooohhh, it's continuous, there's a lot of data there." So figuring out how to do motion whirl exactly in a reasonable amount of time is a tricky thing that we work on all the time.

I: IDM is becoming visible in many fields other than the entertainment industry, like medicine, robotics, telecommunications, geography and architecture. Will the coming future of IDM depend largely on advances in engineering and technology, for example nanotechnology?

R: Nanotechnology? Maybe not so much. Certainly, if you really think about it, the latest chips from Intel and AMD are very much nanotechnology — they are cramming four full-blown processors in a single chip. That has a huge impact — the fact that we can do things in parallel. A lot of what goes on in computer graphics is embedded in what we can do in parallel. This is very good for us. We love the fact that processors are now becoming multi-processors all the time. We buy the latest things from Intel with 8-core or quadcore quad processors. We get 16 processors in the machine. We immediately jump on it and start using it. That's nanotechnology that has a direct impact on us. Other than that, there isn't much nanotechnology. Quantum computing, maybe. There have already been some theoretical uses of quantum computing for computer graphics.

I: Does that mean you have to develop new techniques of software or computational methods?

R: After we had third-generation computer graphics and computer vision, both have adjusted themselves to take advantage of the parallelism of processors. It's a big thing. The artists love it because it makes everything faster. It's not hard to do. With fluid simulation, it's tricky because you need to write optimization parts of the fluid simulation like conjugate gradient, preconditioned conjugate gradient, and other tricks. Once you've done it, then all of a sudden, your fluid simulation comes on.

I: Do you give courses on digital media? Do you have any students?

R: The courses that I give are typically SIGGRAPH courses which are one or two-day courses at annual conferences. People who sign up go there to be educated. In terms of students, for courses, no, but we do have internships. We have interns, master's or PhD interns, people who come here constantly. Right now, I'm working with a student of Tony Chan's, UCLA and there's also a student from a Swedish University working on hair simulation. Yes, there are students. Internships usually last about four months. Either I or another one of our R&D staff is the advisor of these guys.

I: What is your advice to a mathematics student who wants to have a career in your field?

R: If you are in applied math and you know how to program, I suggest you look at the last couple of years of SIGGRAPH's proceedings. SIGGRAPH is the pre-eminent conference on computer graphics and usually there are 90 to 100 papers accepted to the conference each year. Take a look at those and see what kind of mathematics is currently going on. At this symposium, we have Peter Schroeder talking about differential geometry, very hard, very cool stuff, and you will immediately get a sense of the kind of mathematics that is useful to computer graphics and interactive digital media. I'd just focus on that kind of stuff. Going back to what we talked about programming, if you want to work in our industry, you have to know how to program. We don't hire pure math people who just sit and do math. You have to come up with an idea and make a tool that the artists could use, and then the artists... that's the best part of the whole deal, especially in the film industry. There's that lovely feedback when you're working hand in hand with some very creative artistic people. You create something and they will immediately turn around and use it the way you haven't thought to do and they give you a new idea. And you say, "Okay, I'll be taking that back and use it differently." It is so satisfying, it's much better than writing a paper and summing it up in a journal and getting some people saying, "Oh, I saw your paper." This is writing something where people are immediately using it to create brand new things and entertain people and show people new concepts. That immediate feedback, sitting next to the artists and they have great ideas from a whole different perspective than what I can offer. That's the best part of the job. Just that constant creativity from all different sources — we have mathematical creativity, some people read a new paper and say, "Look at this new technique" and then they go and show that to an artist, and he said, "Oh, look, I can do that." It's so fun.

Ron

Unraveling Genes, Understanding Diseases

Ron Shamir made significant contributions to optimization and graph algorithms and is one of the leaders in bioinformatics and computational biology whose pioneering work contributed to the historic completion of the ambitious Human Genome Project in 2003.

He went from Tel Aviv University and the Hebrew University to the University of California at Berkeley for his PhD in operations research. While he is based mainly at Tel Aviv, he has held visiting research positions at top universities and research centers in Seattle, Rehovot, Rutgers and Berkeley. A full professor in the School of Computer Science of Tel Aviv University since 2000, he holds the Raymond and Beverly Sackler Chair in Bioinformatics. He was also the head of the School of Computer Science at Tel Aviv.

He is actively involved in organizational and committee work for international scientific meetings. He has been invited to give lectures at major scientific meetings, research institutes and leading universities throughout the world. He is well-known for his tremendous energy of scholarship in

Shamir

reviewing activities and in serving on the editorial boards of many leading international journals in discrete mathematics, computer science, bioinformatics and computational biology — among them *SIAM Journal on Discrete Mathematics*, *Journal of Computer and System Sciences*, *Journal of Computational Biology*, *IEEE/ACM Transactions on Computational Biology* and *Bioinformatics*.

As the leader of an active and internationally well-known group (ACGT) on algorithms in computational genomics at Tel Aviv, he sets the direction of and contributes extensively to the research that has produced algorithms and software for gene expression analysis, genotype analysis, graph-theoretic tools for modeling biological systems and statistical software for whole genome association studies. Among other things, he continues to contribute to one of the central problems of the post-genomic era, namely, the determination of the function of genes and pathways.

He was an invited speaker at the Institute's program on *Algorithmic Biology: Algorithmic Techniques in Computational Biology* held from 1 June to 31 July 2006, and was interviewed on 18 July 2006. The following is an edited and enhanced version of the transcript of the interview, in which he talked about the excitement of switching from a mathematical realm in theoretical computer science to uncharted virgin territory in bioinformatics and computational molecular biology. Here he also gives us an insightful glimpse of the "brave new world" of modern biological sciences and its impact on human life.

Imprints: You did your PhD at the University of California at Berkeley in optimization. Could you describe the route that took you from operations research to biology?

Ron Shamir: I did my PhD in operations research, but with a very strong computer science tendency. One of my advisors, Ilan Adler, was from the Operations Research Department and the other, Dick Karp, was from computer science. I joined the Department of Computer Science in Tel Aviv a couple of years after that. I worked mainly in the field of optimization problems until around 1990. At that time I was on sabbatical in New Jersey at the DIMACS [Center for Discrete Mathematics and Computer Science] institute, and I did some work on temporal reasoning — in this problem one has to place events as intervals on the time line based on given constraints on the relations between event pairs. At some point, there was a workshop and I presented this work, and the late Gene Lawler was in the crowd, and he told me, "Listen, this is very appropriate for modeling physical mapping of DNA." I didn't know what DNA was or what physical mapping was. Indeed, physical mapping just deals with constraints of intervals along the line, but the intervals are not temporal events but actual DNA blocks. So I started reading biological articles and got excited about this field. It was in the early days of the Human Genome Project, and I found myself part of this new field — in the beginning only partially, but eventually this became my main research interest. The first steps of this transition occurred in Rutgers, New Jersey, but of course, a lot of things happened in the following years.

I: Was there a group doing research in that field?

S: Practically, no. There was one colleague of mine, Haim Wolfson, who was working on structural problems related to biology from the geometric point of view and who got into the field a little earlier. But even so the term "bioinformatics" was not invented yet. We have set up our first formal bioinformatics program in the year 2000 at Tel Aviv University. We came a long way since, with a very strong and active bioinformatics community across the university, including over 15 groups in computer science, life sciences, medicine, physics and engineering.

I: Did it occur to you to continue your work in the United States?

S: No. I got several offers over the years, but never seriously considered accepting them. Israel is my home country, the home of my family, of my nation, and of my mother tongue. Of course, scientifically such offers were tempting, since the research conditions in the United States are better, but science is not everything to me.

I: What is the difference between bioinformatics and computational biology?

S: Actually, I don't think there is a difference. It used to be thought that bioinformatics dealt more with the databases and software, and computational biology used to deal with algorithms. In the past, I used them as two distinct terms, but I do not make the distinction because people are using them interchangeably. We ended up calling our BSc and MSc programs (which should have been called "computational biology") "bioinformatics", because this is the term people are using. Semantically, there is also a technical difference. Bioinformatics is primarily informatics (computer science), and computational biology is primarily biology. But in the way people use these terms, it is the reverse.

I: Structural genomics is often considered as an investigation into the language of genes. Has linguistics or the study of human languages cast any insight into genomics?

S: I don't have much to say about this because I don't know linguistics well enough. The only aspect of it that I am aware of is natural language processing. It's not really linguistics. It deals with collecting the relevant words from large texts like the collection of millions of scientific abstracts and the like. So in that sense, the mechanics of trying to parse scientific texts has been used. I would say that, in general, such approaches were pretty successful, but of course they are not as accurate or provide "clean results" as human investigators would do on the same task. It is good as an initial filter but it lacks human intuition and broad context understanding.

I: Is it correct to say that gene determination is more of a computational and statistical problem rather than a systems problem?

S: It's a mixture. I don't think you can separate them. Gene determination using just statistical or just computational methods has been successful in a limited sense. A few years after the human genome has been completely sequenced, we still do not have the full picture of the genome because our prediction tools are not accurate enough. People have been using additional species to try to get better gene prediction. People have been using the systems approach. I think we are still a few years away from coming up with the ultimate set of genes. This can only be done with integrated methods that use what we will learn from systems biology and comparative biology approaches, and, of course, from classical experimental methods in biology.

I: Are there any general principles which help you to say that there are only 5% of the genes that remain to be determined?

S: Five percent is just a metaphor, not a solid number. It is a rough guesstimate based on extrapolation of what is already known. Five to 10 years ago, we thought that most of the gene regulation occurs at the level of transcription, and now all of a sudden, we have this huge wealth of mRNA, siRNA, microRNA, etc. that changes the picture completely. Who knows what else there is that we are not aware of at this point? For example, there is very exciting work about ultra-conserved regions in the genome that we don't know the structure and function of and there must be a reason that they are so conserved. There is a lot of signal probably hidden at the level of the packing of the DNA and making certain regions more exposed and or accessible for transcription. There's definitely much more in the genome than what we know at this time. There are a lot of exciting revelations waiting for us. That is what makes the field so interesting.

I: In that case, it will depend a lot on the technological advances available at present.

S: Definitely, a lot of it depends on the ingenuity of experimental scientists and on technology development.

I: Has computational work on DNA nucleotides yielded general principles of genome evolution?

S: I think this is probably one of the oldest areas in computational biology because research on evolution and on phylogenetic trees has been an active research area since the '60s. It's a very old area, on the computational biology scale. I do believe we know much more about the selection forces that act on the DNA. At the time of Darwin the belief was that positive selection was probably the dominant force. The general belief now is that most of the evolution is neutral. There are exceptional cases of either positive or negative selection, but neutral evolution is now believed to be more dominant. Of course, there may be surprises in stock for us in the "junk DNA" region, which covers the majority of the genome, and of whose evolutionary role we understand very little now.

I: By neutral, do you mean that it is random?

S: Yes, most of the changes in the DNA do not affect the well-being of the phenotype; most of the mutations are neutral. Occasionally a combination of such mutations will have an effect — even a dramatic effect — on the phenotype.

I: Could it be that anything we don't understand, we say it is "random"?

S: We just don't understand it at this point. We are in the dark but it's definitely not random. Take the occurrence of ultra-conserved regions. It's definitely not random, it's statistically very significant, but we don't understand the role of these regions.

I: If I understand it correctly, much of the DNA of the human genome is considered to be "junk DNA" in the sense that they do not contain recognized functional elements. How can we be sure that indeed they do not contain "recognized functional elements"? Is it possible that they may contain such elements which we are ignorant of?

S: Definitely, yes. There are probably a lot of functional elements that we are not aware of either because we don't have the technologies to identify them or we simply didn't ask the right questions. What happened during the last five to 10 years has shown us that our knowledge is very limited. For example, what happens now with the genome-wide chips is that we used to think that expressions happen only in the regions coding for genes, and now all of a sudden we have evidence showing that a lot of expressions is going on in non-coding regions, contrary to what we thought before. The same thing happens with the binding sites of transcription factors. We used to look at them only in promoter regions. Now when you look at the binding in a genome-wide fashion, using *ad hoc* chips and some of the techniques developed here in Singapore; you see that there is binding all over the genome, very far from known promoters. So definitely there is a lot of function out there that we are not aware of.

I: It may seem daunting for an outsider to go from biology into computational biology or from computer science into biology. From your experience, what is the least painful, if not the best, way to make such a transition?

S: Luckily for the young generation of students, there are already integrated programs. For example, in Tel Aviv University, we have, back in 2000, set up an undergraduate program where students get a full double major in computer science and biology plus a specialization in bioinformatics. So they can really speak both languages. We now also continue a similar program into the graduate level. For the young generation, it is simple. The transition for someone already educated in one of the three fields is indeed not easy. The different disciplines use different languages, both in terms of terminology and in terms of culture. I came from computer science and mathematics. To me a proof is something where you can write "QED" at the end. Once you've done it, the problem is solved. In biology, the notion of a proof is very different. A proof can be re-proved and un-proved. The notion of a definition that we cherish in mathematics does not exist in biology. The terms keep changing their meaning because of new light that is shed on them. A lot of the difficulties in the beginning were created since each area had its own culture and its own way of thinking. There are some cultural barriers in between. Many biologists of the previous generation are not that eager to try to speak the mathematical language. Many mathematicians are not eager to speak biology.

The transition that I went through — going from computer science and mathematics to computational biology — is easier than the transition required for a biologist if he or she does not have any basic training in computation, because first of all, the language of computation is very formal and very hard to pick up in an informal way. You really have to go to classes or digest the textbooks. Also, mathematics is very structured. You cannot learn "B" before you learned "A". Almost everything is very dependent on prior theory, in contrast to biology which is much more "flat". Another advantage for someone making the transition from the computational side is that biologists have wonderful textbooks — they are very clear and full of illustrations. The textbooks in computer science or mathematics are much less friendly; so trying to learn from the literature in mathematics or computer science, if you don't have any prior training, is much harder than trying to learn biology from the books. Even so, it is not easy. It's a difficult process. If you ask, "What is the easiest path?" The easiest path is to be 18 now.

Then you can learn it in an integrated way. There are very good programs both in Europe and in the United States. I don't know if you already have such programs here for undergraduates.

I: We have started to some extent, but we are at quite an elementary stage and still developing.

S: I would suggest — I don't know whether you have considered it or not — even if you don't train the next generation of biologists in computational biology, train them in computation. Have them learn one or two courses on basic principles of computation so that they will have basic knowledge in this "foreign language". It is worth the effort to include it into the biology curriculum. Also, give mathematics and computer science students one or two courses in life sciences, just the principles, so that they will be closer in language. Then, in graduate school, if they decide to go into the other area, they will have an easier start.

I: What about asking computer scientists to write better books?

S: That would be great, but you need people who will be willing to put in the effort. I think it's mainly a marketing issue. There are very good books for software manuals, simply because there are many thousands of people who will buy it. Biology is taught to millions of students. Computer science students are 10-fold or 20-fold fewer; so there is not enough demand. There is not sufficient reward to simplify these texts; it's really hard work to turn something complicated into friendly and simple concepts. And in the end, there is only so much that you can simplify — mathematics is a formal language and a formal discipline.

I: How much benefit has the complete mapping of the human genome yielded to the medical and clinical sciences?

S: Tremendously, and it's only the beginning. For example, we know by now, as a by-product of the Human Genome Project, the causal genes for the majority of the Mendelian diseases. A tremendous amount of knowledge that we now take for granted wasn't there without the Human Genome Project. It has already made a tremendous difference and it will continue to. For example, the hapmap projects which aim to map all the single nucleotide polymorphisms, are still under way. They have already revealed millions of mutations which make the difference between all of us — different features, different life expectancy and so on. Getting such information would have been inconceivable without the reference human genome. We talk about "the human genome" as if it is a unique genome, but it's just a reference. But once you have a reference, you can start zooming in on individual mutations to see how they relate to diseases. I think we are going to reap the benefits of this visionary project for many decades.

I: This sounds like a reductionist view in biology — that everything can be reduced to the genes.

S: Well, not everything. If you can explain 50% of diseases just by looking at the genotype and the other 50% by other causes, this is a great step forward. In five or 10 years, you will be able to have your full genome sequence, for a reasonable cost, and the doctor can tell you, "Look, you don't have to worry about smoking because with your gene combination, it will not make any difference. On the other hand, you should be very worried about your cholesterol or whatever." I don't think the genes are everything, but they account for quite a lot. They will tell us whether someone is more likely to have a particular disease than others, and if a certain lifestyle is going to make a difference for him or her in terms of quality of life. Of course, all this is a blessing but also a curse because the genetic information must be used and not abused. To a large extent, this is not only a thing of the future but is already here today. People have been doing pre-natal tests to identify all kinds of defects, and they will be able to do much more — and more post-natal tests in the future. We can't stop this knowledge, so we'd better use it for the best of our understanding.

I: Maybe in the future we will be able to look at a person's genome and say that he or she will have a stroke at a certain age.

S: I don't think it will be able to tell you that — but it can give you probabilities. You will be able to tell someone that changing the lifestyle will make a big difference in his or her case. Definitely. Eventually, it's all probabilistic. There are relatively few cases of combinatorial fate. It's up to us. The more we know, the more we can control it.

I: Is there a gene that determines the lifespan of an individual?

S: Probably much more than one gene. In mice, scientists found a gene that affects longevity very significantly. We know that this has a lot to do with the shrinking of the telomeres during the life of a creature. It's not a single gene but quite a few genes affect longevity.

I: In that case genetic engineering can lengthen lifespan.

S: That's one of the dreams. I think real genetic engineering in humans is still far ahead, but in principle, we might be able to do so.

I: It seems that in biblical times people had long life spans by modern standards.

S: I think that they just counted differently... They talk about Abraham and Sarah, and Sarah had a baby when she was 90. They probably lived, in modern way of counting, to the age of 30 to 35. Life expectancy at that time was much shorter than it is today.

I: Research in genomics and proteomics usually involves multidisciplinary team effort whereas the tenure system in the universities (at least in the United States) centers round individual achievements. For that reason, a prominent biologist has said that progress in modern biology will come from institutes of research rather than the universities. Do you agree with this viewpoint?

S: No. First of all, the university system is not that blind to joint effort. Credit will be given to several partners in case of joint work. Even in publications from research institutions, there is the first author, the second author and the last; so credit is not spread totally equally. Institutions outside universities have other advantages over the universities because they may be more flexible, and resources can be moved more easily, but I don't think the issue of credit for work is the primary issue. In my own university, we at the Computer Science School have a lot of joint projects with colleagues in the Medical School and the Life Sciences Faculty. If we are successful, then everybody takes credit for the success. The question of how this credit is partitioned is there, but it is not specific to universities. I don't think that the tenure system is an obstacle to interdisciplinary research.

I: For the younger faculty, the emphasis is on single papers rather than joint papers. In that sense, a younger faculty will not want to risk going into another field.

S: But on the other hand, I have some young colleagues in my university, some of whom were my students in the past. In bioinformatics and computational biology, a lot of what they do is joint work. On one hand, it's not single-author papers, but on the other hand, they are involved in more projects, so they have more papers. It balances out. How many research projects can you carry out just by yourself? If you work with other people, you can be involved in more projects because you only do part of each project. I may be naïve about it, but I don't know of cases where this is the main obstacle.

I: You are heavily involved in many projects at the same time. How do you manage to do it?

S: I've been very lucky to have wonderful students. In the last few years, I was also heading the School of Computer Science in addition to running my group and teaching, etc. The secret is really to have wonderful students. You don't need to guide them on every little detail. Once the group has a critical mass, there is joint effort and there is a lot of assistance by the mature students to the younger ones. Also, it's more fun to do more diverse things. I may be doing a little too much, but I have four or five different areas that I try to be active in. As long as each of these areas is exciting to me and as long as I have such wonderful students, I will continue. As my group is quite large, I seldom work on my own. I work with others, mainly students and also colleagues. Students do individual projects, they get individual credit for them and write theses. It's mostly individual work but it's done in a framework of a supportive and unified group.

I: If I may say so, mathematicians are quite notorious in working mainly on their own without getting involved with others.

S: You are right. It's a different culture in computational biology. In my early years, my papers had only one or two authors, but my papers of today may have four or six or sometimes, 10 authors. Part of it is because it is the culture of a different field. Part of it is because the projects are more complex and have more aspects and require more diverse expertise. They are not as deep as pure mathematics projects but they are complex and therefore there are many people and sometimes several groups involved.

I: You hold a number of patents. Do they pertain to the algorithms or the source codes of the software?

S: Only algorithms. The codes are typically protected by copyright, which is a different type of protection. Actually, all these issues are handled by the technology transfer unit at Tel Aviv University. They define what justifies patenting and also copyrighting for software. All the tools we develop are completely free for academic use, and we make an effort to make our tools useful for the academic community. Occasionally, there is interest from the pharmaceutical and biotech industry. In that case, licensing and patenting have to be addressed. But for me, the issue of patenting is of low priority. It is more important for me that our tools will be useful to others. Our software is not as robust and as convenient as commercial software. We don't have the resources to do this, and I also cannot give academic credit to students for doing work that is purely technical, like graphical interfaces. This work is more appropriate for a company. If at some point, some company would like to take the algorithms and the basic software and package them into something fancy, that would be great. We package the software to make it useful for us, and also, we hope, useful for other academic groups and occasionally to pharmaceutical companies.

I: Have you ever gone back to your original field in optimization?

S: I never really left it. I still find it interesting and I still try to find the optimization or graph-theoretic problem behind any computational problem that we address. Over the years, I realized that you have to compromise in terms of elegance what you do in order to be useful to the biologists. Perhaps 99% of the problems in bioinformatics are NP-hard, and only occasionally you can develop approximation algorithms for them.

I: Typically, in spending your time, do you want to dwell on the theoretical aspect of the problem or do you want to find something that works?

S: According to my training, I would, when I just started out, devote 100% of my time to the theoretical aspect of the problem. But if you really want to get new findings in biology, you have to compromise: you will not have time to prove everything rigorously, and you need to develop codes and not just algorithms, because the algorithms by themselves are not useful to the biologists or the medical people. But I still think of many of these problems in terms of optimization. Interestingly, in operations research, there is a strong emphasis on modeling. You have a real-life problem and a big challenge is to formulate it mathematically in a useful way — for example, as an optimization problem in integer programming. In recent years, I realized that in biology, a big and sometimes crucial part of the research is getting to the right problem formulation. In that respect, I am more appreciative now of the emphasis on modeling than I used to be when studying operations research. In addition to optimization, I find myself doing much more statistics than I was trained to, since the bioinformatics area requires it.

I: Is computer simulation done in bioinformatics?

S: Some people do it, particularly for modeling the dynamics of networks. There is also a lot of the use of Monte Carlo methods (I don't know whether you would call them simulation in the strict sense). When it is very difficult to theoretically analyze a particular distribution of outcomes, you can just sample it and see how the results are distributed. It's quite efficient in practice. Of course, there is also the whole field of molecular simulation where you try to study the dynamics of folding and interactions between molecules and which is a huge area that requires tremendous computational resources.

I: From a simple-minded point of view, is it possible to have a model to simulate the rules of combination of the genes by random selection from a large pool of the building blocks of genes?

S: In principle, probably yes, but we are still very, very far away from that.

Albert Nikolaevich

On the Shoulder of a Giant

Albert Nikolaevich Shiryaev is well-known for his important contributions to probability theory, mathematical and applied statistics and financial mathematics, and in particular, to statistical sequential analysis and optimal stochastic control. He has published more than 160 main scientific papers and is the author or co-author of numerous definitive books and monographs in those fields.

He has received numerous prizes and awards such as the Markov Prize, Kolmogorov Prize, Humboldt Research Award, Honorary Fellow of Royal Statistical Society and honorary doctorates from Freiburg University and Amsterdam University. He has served and continues to serve on the editorial boards of many leading journals in probability theory, statistics and mathematical finance. He had been President of the Bernoulli Society, the Actuarial Society of Russia and the Bachelier Finance Society.

Shiryaev

Shiryaev has a long and illustrious career at Moscow State University (as professor since 1970, head of the Probability Department since 1996 and Distinguished Professor since 2003) and at the Steklov Mathematical Institute (head of the Laboratory "Statistics of Stochastic Processes" from 1986 to 2002).

When Shiryaev was at the Institute to give keynote lectures at the Institute's workshop on computational finance, he was interviewed on 26 August 2005. The following is an edited transcript of an unusual interview which gives us some insights into the scientific legacy of the legendary A. N. Kolmogorov (1903–1987), probably the last great universalist of the 20th century. Shiryaev may be considered to be the successor and upholder of the Russian tradition in probability theory established by Kolmogorov.

Imprints: When did your attraction to probability theory begin? How did you choose the topic for your doctoral dissertation?

Albert Nikolaevich Shiryaev: Before I explain how I chose probability as my specialization, maybe I should say how I became a mathematician. When I was in secondary school, I had many interests. I had a strong interest in sports — I played soccer, I did figure skating and for some years I took lessons in ballet. Twice I danced with the ballet group in the Bolshoi Theatre — my instructress worked in the Bolshoi Theatre. At the same time, because of my relatives, I had an interest in rocketry. I lived in Moscow near the famous rocket center and my uncle's father and many others worked in this place. Also I had a strong interest in a diplomatic career and many times I visited the Moscow Institute of International Relations, but finally, of course, I decided to become a mathematician. I participated in different kinds of Olympiad and finally when I got a medal in secondary school, I entered Moscow State University without examination — they accepted me after an interview with the professors.

As a student at the Department of Mathematics and Mechanics, I did not spend much time in mathematics. In some sense, I began to work in mathematics only after five years of university. The reason was very simple. At that time, the ski coach of Moscow University invited me to be a member of the downhill ski team. I was in very good physical condition and very quickly, after three years, I was the champion in Moscow, and in 1957 I even participated in the Second International Winter Universiade in Grenoble. There were 42 participants and I was placed number 4 and number 7 in slalom and giant slalom. For the Russian people, it was quite good because at that time our country was not very well-known in that sport. So, for three years, I spent a lot of time skiing instead of attending lectures. But at the end of the last (fifth) year, I wrote a diploma paper and it was a good piece of work. Then after many conversations, A. N. Kolmogorov finally said to me, "I want to take you as a member of my department at the Steklov Mathematical Institute, but you must select either sports or science."

I was already 23 and not very young for sports; so I decided to stop my active sports life and work in Kolmogorov's department. Kolmogorov gave me many problems and after one year of work, I wrote my first paper with my friend Victor Leonov on the technique of calculation of cumulants. Very soon, Kolmogorov directed me to some applied problems. As a result, I wrote several papers on the quickest detection problem. The first paper was entitled *The Quickest Detection Problem of the Spontaneous*

Effects and this paper became very popular and many people used it and referred to it. D. Siegmund and B. Yakir, who are here, wrote a lot of papers on problems of that type, very often referring to my paper. After two or three years, Kolmogorov said to me, "You already have all you need for your dissertation." So shortly, I wrote my dissertation, and after that, I took the examinations. It was a little bit of an "inverted" situation. Usually, you prepare for and take the examinations in mathematics, languages, philosophy and so on before you do the dissertation, but mine was an "inverse" story.

So I defended my (candidate) dissertation, and in my work for the dissertation, I solved some optimal stopping problems in the Markovian setting. It turned out that stochastic calculus was very important for these problems and I began to work very actively in this direction. I organized special seminars in the Steklov Institute and they were very popular for many years. We published our proceedings and as a result more than 50 of my students defended their theses effectively. They are not "PhD" theses in your sense. In Russia, we have two dissertations — the "PhD" dissertation and then the "Doctor of Science" dissertation. Generally, out of 10 who wrote the first dissertation, only one will go on to write the Doctor's dissertation.

As a result, I published a book on optimal stopping rules — two editions in Russian, and one translation into English for Springer. I also wrote with my pupil Robert Liptser books on stochastic processes and our main interest was in nonlinear filtering. At that time, we realized the importance of the theory of martingales, and we worked actively in this area. As a result, we also wrote a little book on the theory of martingales and with my French co-author Jean Jacod the book *Limit Theorems for Stochastic Processes*. I began to work in the Steklov Institute in 1957 and I am still its member.

I: Were you also a member of the Mathematics Department of Moscow University?

S: Yes. Kolmogorov attracted me to Moscow University simply for his lectures on probability theory. He was in both places — Moscow University and the Steklov Institute. He was Chair of the Department of Probability in Moscow University. After him, the Chair was B. Gnedenko. Now I am the Chair of that department. It is a very big department. Every year, we accept more than 50 students for specializing in probability, and we have two groups of students — one for probability theory and the other for actuarial and financial mathematics. In 1994, I began to work in financial mathematics and was probably the first to give lectures on financial mathematics in Moscow University. I wrote a big book, published here in Singapore, on *Essentials of Stochastic Finance*. This was reprinted five times and is popular. Recently, the second Russian edition has been published, and World Scientific has asked me about the second English edition, but I have no time because I am writing several books, one book with my colleague from Denmark Goran Peskir on optimal stopping free-boundary problems, another book with Ole E. Barndorff-Nielsen on change of time and change of measures, which will be published by World Scientific.

I: Was your Russian book on financial mathematics the first book published on the subject in Russia?

S: It was first published in English, and it was the first book on financial mathematics published in Russia. Even the great Russian newspaper *Izvestia* published good reviews of the book, and they said that it was important for the development of the Russian economy. At the same time that I was working on the book, I was involved with several publications and books about Kolmogorov. Before the death of Kolmogorov in 1987, we published three volumes on the selected works of Kolmogorov. I was involved in it and I felt that I had to do it. Now we plan to publish six volumes of his selected works: the first volume embraces the papers by Kolmogorov in mathematics and mechanics, the second on probability theory and mathematical statistics, the third on the theory of information and the theory of algorithms, the fourth about mathematics and mathematicians, and we have plans to publish also Volumes 5 and 6.

I: Are the complete works of Kolmogorov published?

S: Not all. Of course, not. In fact, the archives of Kolmogorov belong to me in some sense. Kolmogorov's widow wrote in a testimony that all the archives of Kolmogorov's work belong to me. I must say the following. Two years ago, in 2003, we had a very big conference dedicated to the centennial of the birth of Kolmogorov. I was essentially involved in its organization, and we published before this conference three books related to Kolmogorov. The first volume consists of two parts: the first part is the biography of Kolmogorov written by me (more than 200 pages) and the second part is a bibliography of his work (mathematical papers, papers in encyclopedias, textbooks and papers for secondary schools and so forth). The second volume is a big volume on the correspondence between Kolmogorov and Paul Alexandrov, the creator of general topology. They were friends and had a very interesting correspondence with a lot of mathematical visions. Finally, the third volume is very interesting — it's the diary of Kolmogorov. Practically nobody had seen it before. I found the diary in Kolmogorov's country house. So we published it and now I want to propose to World Scientific to publish an English translation. The diary is very interesting and unusual — he began to write the diary when he was 40. At the beginning, Kolmogorov wrote the following: that he dedicated his diary to his 80th anniversary in the hope that he will understand what he wrote in his forties. In this book, you will find a lot of interesting pages. One page is of the following type — "what I must do to be a great person". Of course, he wrote this ironically. I want to show you an interesting page — his future mathematical plans.

I: How many years did the diary cover?

S: Not too much — two to three years. Here it is very detailed. But later he wrote, not very periodically — his plans on what he should do from 44 to 53, from 54 to 63, etc.

I: Did he follow his plans?

S: Yes, surprisingly, surprisingly. For example, everybody was surprised why he practically stopped working in mathematics after 60 when he started working for secondary schools. But he wrote this down and he planned it. He "predicted" that he should work in this area.

I: Did he stop working completely in mathematics after 60?

S: Of course, he worked in mathematics, but he spent a lot of time writing textbooks on algebra and geometry for secondary schools. He organized a special school and a journal for students of 15–17 years of age and with special abilities in mathematics and physics. He worked and gave classes like the usual teachers.

I: Was it in Moscow or all over Russia?

S: It was in Moscow but the boys and girls were from different cities in Russia. Even students from Moscow may not be selected. They were specially selected by local or All-Union Olympiads.

I: Was Kolmogorov a good lecturer in schools?

S: It's a bit difficult to say. In fact, his manner of speaking was very fast. Very often he jumped and omitted intermediate steps. It was his usual style, and many people said that it was very difficult to follow him.

I: It must be very interesting working with Kolmogorov.

S: It was, of course, very interesting but it was not simple. It was definitely clear that if you did not work very successfully, he would lose interest in you. In some sense, you should be at a good level and have initiative and as a result we had to spend a lot of time mathematically when we were young.

I: Were you very close to Kolmogorov personally?

S: Of course, I know Kolmogorov personally very well. When I began to work at the Steklov Mathematical Institute together with my friend Victor Leonov, Kolmogorov asked us to be his informal secretaries. We attended his lectures and wrote lecture notes for students. As a result, we lived practically two days every week in his country house. We skied with him, and later, I had a car and we visited many Russian cities because Kolmogorov had a vast knowledge of Russian icon art. He knew a lot about Russian churches and the details about their construction and so forth. Communication with him was not very simple because often you have the feeling that there was some kind of screen between you and him. You always feel that in front of you is a brain which works continuously and it was amazing that at the same time he has the ability to think about different topics.

I: You must have felt tense.

S: Yes, at all times you would feel the tension. He was so "non-trivial" that you could not say some trivial thing. He knew music very well — also literature, poetry, archaeology, history, geography. He had a fantastic memory, especially for geography, history and so on. You know that his beginning was very unusual — his mother died two hours after he was born. His father was killed during the civil war, and essentially, his mother's sister brought him up. When he was five years old, he made a lot of unusual observations. He discovered, for example, that $1 + 3 = 2^2$, $1 + 3 + 5 = 3^2$ and so forth. I asked him how he understood those things. It turned out that his solution was purely geometric. He also solved the following problem when he was only five years old: Suppose you have a button. You can fix it to the coat if the thread goes through at least two button holes. The question is: in how many ways? He arrived at an absolutely correct answer. He was only five years old. From the beginning, he had a strong mathematical ability.

I: Nobody taught him?

S: No, he did it all by himself. When he was 12 or 14, he studied mathematics at a very high level, reading mathematical texts in encyclopedias and trying to reconstruct the proofs. So he began very, very early to do mathematics.

I: Nowadays, mathematics is so wide and highly specialized and yet Kolmogorov did so many fields.

S: Two years ago, we had a conference with the title of *Kolmogorov and Contemporary Mathematics*. We had six sections in which Kolmogorov worked: dynamical systems and ergodic theory, theory of functions and functional analysis, theory of probability and mathematical statistics, mathematical logic and complexity, turbulence and hydrodynamics, geometry and topology. He wrote a lot of papers in all these areas and he essentially created many fields. He is the father of modern probability theory. Even the topological notions in cohomology were introduced by him. In turbulence, there is a famous law called the "two-thirds law" of the type of Newton's laws, and it is his contribution. He introduced the notion of complexity which gives the possibility of applying probability even to non-probabilistic objects. The notion of complexity is the crucial clue. I remember that before the organization of the Kolmogorov conference, I was thinking about getting money for the organization. At one point, I asked Microsoft for money and they gave the money, saying, "Yes, Kolmogorov! It's very important in complexity, mathematical logic, computing…" I also asked money from Boeing and they also gave us money (because of turbulence of Kolmogorov).

I: Do you think that in the future somebody can be like Kolmogorov, versatile in so many fields and with so much impact?

S: It's hard to say. In some sense, it is difficult to predict that we will have a person of the following type. Let's look at the "encyclopedic mathematicians" [universalists] — Poincaré, Hilbert, von Neumann, Kolmogorov. It's practically very difficult to give another name.

I: What about Norbert Wiener?

S: He was broad, but I think Kolmogorov worked in many, many different areas. We know Wiener's work in filtration, interpolation but Kolmogorov did that before him. Wiener wrote in his books *I am a Mathematician* and *Ex-Prodigy* that Kolmogorov discovered it a little bit earlier. Wiener, of course, did something in probability — he introduced Wiener measures and the properties of the so-called Wiener trajectories, but in some sense they were particular cases. Kolmogorov was very wide and he was great in the creation of new concepts like complexity in mathematics. Probability space, conditional probability and expectation belong to him. I wrote that if we take a Russian mathematical encyclopedia, we find Kolmogorov axioms, K duality, K integral, K criterion, K inequality, K space, K equation, K-Smirnov criterion, K-Chapman equations. If you take any encyclopedia on probability and mathematical statistics, you will find Kolmogorov axiomatization, K self-similarity, K law of two-thirds, K criterion, K matrix, K model, K distribution, K statistic, K law of five-thirds, K self-similarity, K spectral theory.

I: Did Kolmogorov ever meet von Neumann?

S: Yes, von Neumann gave the opening talk at the Amsterdam Mathematical Congress and Kolmogorov gave the last talk. They had a very short conversation. As I understood, it was not a very long discussion. As for Wiener, Kolmogorov met him in Moscow, but with Wiener it was a very strange story. When Wiener arrived, he made a call to Kolmogorov and said, "I want to meet and have a talk with you."

Kolmogorov said, “Please come tomorrow at six o’clock.” And Wiener came but at six o’clock in the morning. [*Laughs*] For us it is clear that it must be 6 pm.

I: Could the Kolmogorov “phenomenon” be a result of the Russian system and environment?

S: It is difficult to say. In some sense, he was a genius from the beginning. When he was a student of Moscow University, the mathematical school of Lusin flowered and many well-known mathematicians appeared at that time — Lusin, Khinchin, Kolmogorov, Novikov, Petrovskii and many others. It was a special period when the Moscow mathematical school understood that the methods of the theory of functions are very important. In some sense, Kolmogorov said that his success in the creation of probability theory was based on the understanding that the theory of functions and theory of measures play a very important role. Kolmogorov was not a member of the Communist Party but the high-ranking leaders of the Party realized how great Kolmogorov was.

I: The Soviet government appreciated and understood the value of Kolmogorov…

S: Yes, exactly. In the Soviet Union, the highest decoration for anyone is the Order of Lenin. It was awarded to Kolmogorov seven times for his contributions and work in mathematics. There is a famous international story: in 1940 when we had a person called Lysenko who wanted to close genetics down. But at that time, Kolmogorov wrote a paper about the brilliant confirmation of Mendel’s laws. Politically, it was very dangerous but nobody arrested him. In the beginning of the Second World War, Stalin asked Kolmogorov to begin work — not military work, but work related to the defense force. By the way, two years ago, there was a conference entitled *Mathematics and War*. They invited me and I wrote a paper about the defense work of Kolmogorov during the Second World War.

I: It was not classified work? The Russian government allowed it to be published?

S: It was a mathematical problem. In the beginning of the war, we had in Russia many light small planes. Suppose we use these planes for bombing and it is necessary to predict where the bombs will land. It depends on the velocity and so on. It was necessary to create ballistic tables for bombing. Kolmogorov did it by himself and he discovered the following interesting phenomenon. Suppose you have a bridge and want to destroy the bridge. Usually, we would want to hit the center of the bridge. But Kolmogorov discovered that, in fact, it was necessary to create an “artificial” bombing. You would try to aim at one point, but he said, “No, it’s not correct. Sometimes it is necessary to try to hit here or try to hit there.” In other words, it is necessary to create an “artificial deviation”. It was the beginning of much work of this type and he created the tools for this work.

I: It seems that the Russian system of education is very successful in developing problem-solving skills. What is it due to?

S: In some sense, it is true. The basic reason is that Russian mathematics has a very good tradition in the secondary school education and in the universities. We had a lot of great mathematicians who created many different schools of mathematics. Kolmogorov created the school of probability theory, Petrovskii created the school of differential equations, Novikov and Markov created the school of algebra and mathematical logic, Pontryagin created the school of continuous group theory and after that, he worked in the theory of optimization (the Pontryagin maximum principle). And simply, we have great people who created scientific schools and they were related to the education in the universities.

It reminds me of the following. At the end of the Second World War, Lysenko or Stalin (I don't know exactly who) said that science is the enemy of randomness in the sense that science is trying to make order in everything. But the representative of the philosophy school began to attack probability theory, saying that probability theory investigated the notion of independence, but everything in the world is related, and so the notion of independence is nonsense. They then said that Kolmogorov's Department of Probability is dealing in idealism. Kolmogorov was invited to the conference, and there was a discussion about independence and randomness. And Kolmorogov told them, "Let's take the government lottery. The randomness that you will win is guaranteed by government. Suppose that it is not true. Then it means that the government creates unfair lotteries."

I: Your faculty is called the "Faculty of Mechanics and Mathematics". For us, mechanics and mathematics seem to be a strange combination...

S: It's theoretical mechanics. We have in this faculty two sections. One section is in mathematics, and it is clear what they are doing. The other section is in mechanics; they are doing turbulence, hydrodynamics, elasticity — in some sense, partial differential equations with applied aspects. They investigate what the form of the airplane should be, how it depends on the velocity and so on, but by using mathematical methods. They have some part of the engineering aspects, but it's mainly the theory.

I: Do you consider yourself to be an applied probabilist now?

S: No, of course, not. I remember that at a banquet after defending my doctor's dissertation, several people proposed toasts. One person said that Shiryaev was a probabilist, another said that he was a statistician, another said he worked in applied probability. But Kolmogorov said, "We are mathematicians, and if you are a good mathematician, you should be able to solve any problem — theoretical problems, applied problems and so on." Now I am doing financial mathematics, but I don't work directly in the bank or for the banks. Simply, we understood that financial mathematics and financial engineering give rise to a lot of new theoretical problems, and we are trying to solve them. In the coming conference on Monday and Tuesday, the talks will be exactly about this.

I: But now you are more interested in financial mathematics.

S: Not exactly. I think it will be very bad if I will concentrate only in financial mathematics because, first of all, around me there are a lot of students, and as head of the Department of Probability, I should have a good orientation in many different theoretical aspects. I cannot give them only problems in financial mathematics because I must think about developing probability theory as a science and developing mathematical statistics as a science. From some point of view, financial mathematics is now very attractive because of its many new problems and more job opportunities. It is also necessary to mention not only financial mathematics but also actuarial or insurance science. In Russia, I was the

President of the Actuarial Society for four years, and so we began to work in this direction. It's true that our students were able to get good salaries after university. That is, of course, important, but in some sense it is a pity that a lot of our good students leave Russia to continue their education mainly in the United States and England. Many of them got jobs in the United States and other countries.

I: Is your department doing anything to retain your talents and encourage them to stay in Russia?

S: It's a very difficult question. I know that several people have returned to Russia, but look, you have young people going to the United States for their dissertation and it is a period when they begin to have families, children, houses, and life is life. So they continue to stay in those places. But I know several cases, not young people, who retired in Russia. But now, it is not easy to get good positions in Russia. For example, our Steklov Mathematical Institute is very small. It's a good institute; in some sense, it's like the Institute for Advanced Study in Princeton. We are doing theoretical work and we are proud to be members of this Institute.

I: How many members are there in the Steklov Institute?

S: We have 12 departments in the institute and about 120 research members. They are permanent members. We have practically no visitors. I worked in the Steklov Institute all my life and I am very happy. The Institute belongs to the Academy of Science and if we ask for a new position for a good young person, usually we get it.

I: What do you think of the future of mathematics in Russia?

S: Of course, we want to continue the good Russian tradition in mathematics. I want to say that our administration of the Academy of Science, for example, is trying to do it. Who is the President of the Russian Academy of Science? Academician Yu. Osipov who is a mathematician. Who is Director of the Steklov Mathematical Institute? Academician V. Kozlov who is Vice-President of the Academy of Science. Who is the Rector of Moscow State University? Academician V. Sadovnichy who is also a mathematician. They have a lot of power and they are trying to preserve the tradition, not only for mathematics but in Russian science. As a result, we have a good administrative group for mathematics. Of course, they are doing a lot of things for many other fields, but I think that it is a positive point for preserving the good Russian tradition in mathematics. Also, many academic people paid a lot of attention to education in secondary school and in university. This is to keep the good Russian tradition. That is why, for example, in the '40s and '50s, mathematics in Russia was good. For example, Kolmogorov worked purely in research in the Academy of Science and at the same time worked for education in Moscow University. Good scientists in research at the same time gave lectures and seminars at the universities. As a result, students get a good opportunity in understanding in which direction it is necessary to work. There is consolidation and interplay between the Academy of Science and education. This is very important and, in some sense, it increases the possibility of keeping our good Russian tradition in mathematics.

David O.

Change-Point, a Consequential Analysis

David O. Siegmund is widely acclaimed for his fundamental contributions to the theory of optimal stopping time in sequential analysis and for his recent work on the application of analysis to genomics. He is well-known for his philosophical delight and mathematical ability in commuting between the theoretical heights of probability theory and the murky depths of statistical applications.

He taught for about eight years at Columbia University, where he obtained his doctorate under the supervision of Herbert Robbins. Since 1976, he has been at Stanford University, where he was Chair of the Statistics Department twice, served as Associate Dean of the School of Humanities and Sciences and is now the John T. and Sigrid Banks Professor. He has been a visitor to The Hebrew University, University of Heidelberg, University of Cambridge and Oxford University. He was at NUS in 2005 as the first Saw Swee Hock Professor of Statistics.

He has been invited to give lectures at major scientific meetings; in particular, the Wald Lectures, Hotelling Lectures at the University of North Carolina, Taiwan National Science Council Lecture, and

Siegmund

Bahadur Lectures at the University of Chicago. Among the many awards he received are the Guggenheim Fellowship, Humboldt Prize, Wilks Medal and membership of the American Academy of Arts and Sciences and of the National Academy of Sciences, USA. He has served extensively on professional committees in the United States. He has also been on the editorial boards of leading journals, such as the *Annals of Statistics* and the *Annals of Probability*. He was President of the Bernoulli Society and of the Institute of Mathematical Statistics. His numerous papers deal with theoretical questions in probability theory as well as concrete applications concerning clinical trials and gene mapping. He wrote two books (the first jointly with Y. S. Chow and H. Robbins) which are now classics in sequential analysis.

David Siegmund's long association with NUS dates back to the 1980s (as external examiner for the University of Singapore) and continues as a founding member of the Scientific Advisory Board of the University's Institute for Mathematical Sciences since 2001. He was interviewed during his visit to the Department of Statistics and Applied Probability (DSAP) from October to December 2005. The following is an edited and revised transcript of this interview in which he talks passionately about his early attraction to mathematics, his subsequent search for the relevance of the mathematical sciences and a calling which he finds fascinating and challenging in theory and application. Here he also shares with us his rich experience in research and administration.

Imprints: Were you already fascinated by statistical mathematics in your school days? Were your school teachers instrumental in attracting you to statistics?

David O. Siegmund: The answer to the first part is clearly "no". In my school days, I had one mathematics teacher whom I liked very much, but at that time I was more interested in the foundations of mathematics. I found a book describing Cantor's set theory, the cardinality of infinite sets, the non-denumerability of the real numbers, etc. I thought that was a beautiful subject. I did have a university teacher who was instrumental in my attraction to statistics. In some sense, I became interested in statistics because I became disenchanted with the way mathematics in the 20th century had divorced itself from science. I took up an interest in this science and that science, shopping around, and at one point tried the social sciences. After deciding that none of these was exactly right for me, but with an interest in the social sciences, I was drawn to statistics as an area of mathematics closely related to the social sciences. Ironically, I have never done anything specifically related to the social sciences since then, but it did play a role in helping me find the field of statistics.

I: Were you more interested in applications than theory?

S: I've always been interested in theory. At heart, I would love to be a pure mathematician. At the same time I always wanted problems that seem to be related to some kind of applications, but they certainly don't have to be applied problems in the sense that working applied statisticians would recognize them as applied problems.

I: How did you get interested in sequential analysis?

S: During my last year in the university, I took a course that involved reading Volume 1 of Feller's book on probability theory (at that time there was only one volume, now there are two), and I thought that the chapter on gambler's ruin was both fascinating and mysterious. The problems were fascinating, and while setting up difference equations was very natural, pulling solutions out of the air, as it seemed to me at that time, was very mysterious. In my first year at graduate school, I took a course in sequential analysis from Herbert Robbins and found the same problems were considered there from a completely different point of view. The methods of solution seemed more satisfying, and the connections to statistical applications added to my interest in the classical problem of gambler's ruin. Since then I have been interested in sequential analysis.

I: Was your PhD thesis on a topic in sequential analysis?

S: It was — on optimal stopping theory. One of the first things I read on my own during the first summer I was a graduate student was the chapter in Doob's book, *Stochastic Processes*, on martingale theory. I thought that it was the most beautiful mathematics I had seen up to that time, and it was naturally related to optimal stopping theory. Conceivably, I had the motivation from sequential analysis at the time but I don't recall. I think I just wanted to learn stochastic processes and that was one chapter that particularly appealed to me. Since my PhD thesis advisor, Herbert Robbins, was interested in optimal stopping theory, and it was naturally related to martingale theory, it was the subject for me.

I: That was at Columbia?

S: Yes, that was at Columbia.

I: Is Columbia near your home town?

S: No, I grew up in St Louis which is right in the middle of the United States. I started to think about Columbia because my wife was interested in going to the Columbia School of Social Work, probably the best known school of social work in the United States. When I mentioned this to Paul Minton, who advised me as an undergraduate, he became excited and said, "Oh, Herbert Robbins, now at Columbia, would be a wonderful advisor. He is very creative. You would love to work with him." So my wife's interests and my interests seem to coincide, and we went off to New York.

I: Robbins was originally a topologist?

S: He wrote his PhD thesis at Harvard in topology, but then before he really developed as a topologist, he was led during the war to problems of operations analysis. After the war, he was invited to become a Professor of Statistics even though he had never taken a statistics course in his life.

I: In your scientific career, you have moved between Columbia and Stanford. What made you decide on Stanford as your eventual choice?

S: From a professional point of view, I found different advantages at Columbia and at Stanford, but my wife was an unequivocal spokesperson, on behalf of our children too, in favor of Stanford. I think she was completely correct — it is a much nicer place to live in than New York City. The scientific advantages became clear to me later on, though early in my career I liked very much to be in Columbia. But Columbia is not as strong a scientific university as Stanford is, and the statistical applications one naturally comes across in New York outside the university have to do with the financial community, the legal community and so forth. Those were interesting but I did not naturally gravitate to them the way I gravitate to some of the scientific things at Stanford. And Stanford's Statistics Department was larger and certainly, on average, a better department. So that seems to have been a good choice in the long run.

I: I believe you were at Columbia for quite a while.

S: I went there for three years as a graduate student and beginning assistant professor, with a one year hiatus at Purdue University, where Y. S. Chow was on the regular faculty and Robbins and Aryeh Dvoretzky were visitors. After two years at Stanford as an assistant professor, I went back to Columbia for seven years. But, since 1976, I have been at Stanford.

I: Do you consider your work from sequential analysis to change-point analysis a natural development of your scientific interests? Could you tell us something about the origin of change-point analysis?

S: It was certainly a natural step. I really didn't know much about change-point analysis; but Bruce Macdonald, who headed the statistics section of the Office of Naval Research asked me to give a seminar in Washington, because he thought some of my research might have applications to change-point analysis. I went there with a few of my own thoughts, but in ignorance of the existing literature. Some of the questions asked by the audience and some of the references they mentioned made me aware that there was this field of change-point analysis. I realized that it was indeed closely related to what I had been doing in sequential analysis and that it was quite interesting. In a sense, change-point analysis began with quality control at Bell Telephone Laboratories in the 1920s and 1930s, but the real breakthrough, which ushered in the modern period, involved a couple of papers by E. S. Page, a British statistician, in 1954 and 1955 when he introduced the CUSUM test as a means of quality control. He didn't understand the relationship of the CUSUM test with the likelihood ratio test of statistics. That understanding came later, but since that period in the 1950s the subject has grown quite a bit. Initially, it was the result of the seminar questions that I didn't know the answers to, but then later the very rich theory and applications, that have held my interest.

I: You mentioned change-point analysis as a quality control thing. Was it empirically motivated?

S: Yes. The conceptual scheme is that we have a process, some kind of industrial process, that produces items in a complicated way that amounts to a black box. We can't look inside the black box to see if it is operating correctly. What we can do is to make measurements on the products to infer indirectly if it is operating correctly. The change-point philosophy was that you are careful in the beginning when you set things up, and the black box will initially operate correctly. Then after a while, someone gets careless or machinery wears out, and there is a change in the product, and you have to spot that change and then make adjustments to the system so that it starts operating correctly again.

I: Is there a theoretical foundation for this?

S: There certainly is a mathematical foundation. From the point of view of applications, there is always a debate whether a particular model is the best model that you can use. There are models where changes occur instantaneously by a discrete amount and others where changes occur gradually. There is a debate on which kinds of models are better. In spite of a certain level of implausibility, by and large the model that posits abrupt changes is very successful.

I: Do I understand that there are many change-point models?

S: Yes. There is no canonical problem. A problem has a certain structure to it but there is not a single mathematical formulation. In fact, I am sometimes at a loss for terminology. The term "change-point" is embedded in people's minds, but there are many problems with the same essential mathematical structure that don't really fit the change-point idea. So I sometimes use the phrase "change-point-like problems" to convey the idea that we are doing something related to change-point problems but it's not what you would automatically expect.

I: How do you choose the model to use when you are doing change-point analysis?

S: I don't think the answer is any different from any other statistical analysis. One typically starts with the simplest possible model that seems to capture some of the conceptual features of the problem, and then starts adding complications, sometimes called "bells and whistles", to make the model more satisfactory in a quantitative sense, although there is always the desire to keep the things as simple as possible conceptually. There's a famous statement of Einstein to the effect that a theory should be as simple as possible but no simpler. It's the same thing in choosing a model.

I: Do you know whether change-point analysis has been applied to data in the social sciences or even in the historical studies of cultures or linguistics?

S: There is a simple answer to the question, which is "yes", but I can't very effectively describe these applications. There are some in economics and finance, which in fact was the origin of some of the early applications of change-point analysis. In finance, for example, my colleague at Stanford, T. L. Lai has developed quite sophisticated change-point models that can lead to different investment strategies from time to time. I also occasionally get sent a paper or am asked to comment on a paper in the social sciences that has a change-point aspect to it. I usually forget these pretty quickly, so I don't really feel comfortable trying to discuss them in detail. But, for example, I do recall some research concerned with learning theory that asked the question whether learning, say simple skills in elementary school, should be thought of as something that proceeds by occasional dramatic improvements, where testing would indicate that someone hasn't learned anything but then seems to learn overnight, or alternatively that tomorrow we will be a little better than we are today and the next day we will again be slightly better. The learning theorist was trying to build a theory suggesting that progress appears to be rather abrupt, which would be consistent with a change-point model.

I: In history, for example, there are events which are marked by changes which can be thought of as change-points.

S: Right, there's certainly some of that motivation for applications in economics. People ask whether certain policy issues actually lead to changes in behavior or changes in economic conditions or whether certain external shocks to the system lead to a dramatic change or lead effectively to no changes at all. Conceptually, that kind of issue has been a part of some economic thinking.

I: Is it possible to use change-point analysis to make history more quantitative?

S: I don't know. Historians try to use surveys and quantitative methods more and more. It would be interesting to know what kind of change-point models there might be. One problem that is an interesting conceptual application of change-points (and has an historical aspect to it) is the set of data examined by many people, which involves fatal accidents in British coal mines. For about 150 years, the British Coal Mining Board recorded accidents, and kept very clear records. Every accident that involved the deaths of more than 10 miners was recorded. During the period around 1890, there were royal commissions that studied the problem and made recommendations for how mining practices should be changed to make them safer. People naturally wanted to know whether this had an impact. Indeed, the rate of accidents dropped quite precipitously, or the average time between accidents increased quite sharply around 1891–1892, during the time that these activities took place. One presumes that this was a response to changes recommended by the commissions, which involve things like, if I recall, using a different kind of explosives, one that is less flammable, using water to wash down the interior of a mine, in particular, trying to get coal dust out of the atmosphere.

I: Is change-point analysis extensively used nowadays?

S: It is certainly widely used in the sense that you can find versions of change-point analysis in many, many different scientific contexts. Within those contexts, it's fairly specialized. For example, in drilling to find oil one wants to know something about the density of rocks through which one is drilling and in particular changes in density reflecting changes in the mineral composition of strata encountered during the drilling process. Change-point analysis of magnetic resonance image data is an approach to this problem that has a somewhat different flavor from most other applications I'm familiar with. Change-point analysis of DNA sequence data has recently become popular in some problems of molecular biology.

I: What about to evolutionary biology?

S: I guess there should be, but I've never looked at the data, and I don't know whether anybody has actually tried to formalize such a model. Certainly there is this ongoing debate about the hypothesis of Stephen Jay Gould of a punctuated equilibrium, that evolution doesn't proceed by small incremental changes as people more or less inferred from Darwin, but exists in a steady state without much in the way of changes and followed by a large number of changes occurring rather rapidly. I think this is a rather natural reaction to reflecting about the role of the environment in evolution, because we know that there are things like ice ages, meteors hitting the earth and volcanoes that have drastic impact on the environment leading to dramatic changes in, say, the average temperature of the earth and the seas. So it's natural to think that those changes, if they occur quickly, must lead to rapid changes in flora and fauna as well. But I don't know if anybody has actually tried to build a model and address the issues quantitatively. It would certainly be interesting, but it is also likely that the data are not sufficient, since this involves going a long way back in the history of the earth to find appropriate data. More modest questions of an evolutionary nature involve change-point analysis of DNA sequence data to identify, for example, places where mutations occur more frequently than the overall background rate.

I: Am I right to understand that the identification of a gene is a change-point problem in DNA analysis?

S: It certainly can be viewed that way. I would say it is helpful to view it that way, although most people involved in gene mapping, which is the area I'm primarily interested in now, do not share my view. I think they are missing something. With the advances made in technology that allow one to genotype markers closer and closer together, the change-point aspect of the problem will become more apparent. Historically, there were very few markers distributed across the genome. For the last 10 years, in human genetics it has been customary to use on the order of 300 to 500 markers. Even at that level of resolution, the change-point aspect of the problem is not quite so apparent; but if the resolution should ever become what would be implied by having thousands of markers, which one can easily imagine, then the change-point viewpoint will increase in importance.

I: Is the problem of gene determination in the human genome completely solved?

S: No, it's one of those problems where progress seems very rapid, but then one realizes that there are still many more problems. With each step that we can take, we become more ambitious. Not so long ago, one didn't try to map genes except for very simple diseases where there was one gene involved and the gene literally over-ruled almost anything in the environment to determine the phenotype of the individual. Now, one is interested in what are referred to as complex diseases or quantitative traits that involve both the genotype, of possibly multiple genes, and the environment, which also may interact. These problems are much more difficult. As I said, at each stage when we think we can tackle more ambitious problems, we realize that the number of problems that appear to be solvable has actually grown and not shrunk.

I: What about the total number of genes in the human genome? Is that settled?

S: I'm skeptical, but the answer commonly given is about 30,000. Only a few years ago, people were guessing 100,000. I would guess 30,000 is closer. But that ignores features that have only recently been recognized as important. One of these is what is referred to as "alternative splicing" so that a single gene, depending on how the pieces of the gene work together, can produce more than one protein. The mantra of molecular biology 50 years ago was "one gene, one protein". Since there are many proteins, one had the idea that there must be many genes. Now, it appears that the number of human genes is much fewer but the number of proteins is still very large. So the basic problem doesn't change simply because we now say there are fewer genes. There is still a large number of functions that are incompletely understood.

I: Has any work with the computer led to theoretical insight in your research work?

S: I don't have a very good answer for that. I think that the computer is so much an intrinsic part of my research that it's hard to say what is an insight based on something I've done on the computer or some other kind of insight. It's very easy to say that the computer helps eliminate paths of research leading to dead ends and reinforces fruitful pathways. But working out detailed examples with paper and pencil is the more old-fashioned way to the same result. To some extent, I'm an old-fashioned person. What insights I've had come from piling up individual cases and trying to find the general pattern. I'm very envious of people who seem to get insights without compiling lots of special cases and who seem not to need to do the calculations until they already know what it is they want to calculate. In my case, most calculations are wasted. There's always a pile of papers on my desk. I cover them with scribbles and throw them away very quickly. The computer is helpful in saving some of those efforts in certain cases. Another very important consequence for statistical analysis is that the computer redefines what one means by a solution to a problem. There are still things computers can't do, but basically a problem is solved once it's reduced to something computers can do. Of course, even then, that is not a completely clear answer because what a computer can do in one person's hands is much more than what it can do in my hands. I have the good fortune to work with many good graduate students and younger colleagues, all of whom know computing better than I do. Often they will keep me from spending too much time in blind alleys by doing some computing for me.

I: Is there any software for the application of change-point analysis?

S: People do develop software for change-point analysis. I don't know of any commercial or large-scale programs largely because I don't use such programs on a day to day basis. I'm very poor at using other people's software, so when I want to do some computing I usually write primitive programs of my own. I have seen software that advertises the ability to do change-point analysis but I have never looked at it carefully to decide whether it is the right way or the way I would do it. Software development is a valuable activity, but it's not for someone of my primitive computing skills.

I: How often do you interact with clinicians and medical practitioners?

S: Here we have an issue of the definition of "interact". If interact means to sit down in an office face to face and have an in-depth discussion of a problem, the answer is "not very often", a couple of times a year. If it means to have a more superficial discussion trying to see whether we have common ground for deeper collaboration, then it's certainly much more often. Many of these discussions, I think, don't lead directly to that collaboration, but I find them very useful nevertheless in trying to formulate problems. Often my formulations are fairly theoretical, so I don't try to propose my research as an immediately practical solution; but I find these discussions a very useful conceptual bridge to finding an interesting research problem. If you broaden the definition more to mean reading articles in medical or genetics journals that don't themselves have completely satisfactory solutions to their statistical problems, I would say I spend a great deal of time doing that. That may be one of my primary sources of stimulation in finding problems. When I was much younger, I read the mathematics, statistics and probability literature to improve my techniques in solving problems that were already formulated. Now I depend on other people to tell me if there is an interesting new mathematical or statistical technique, and what I am really more interested in finding out is if there are new scientific problems that are to my taste, which is somewhat idiosyncratic. That may not be what people mean by interaction, but it's interaction at a distance, by the printed page, and I do that a great deal.

I: Do you interact through meetings or conferences?

S: Certainly. Each year, I attend a few statistics meetings and a few genetics meetings. The main reason for going to the genetics meetings is to find out the way the science is going and to try to infer what are interesting statistical problems from what people are taking about. These can be problems that they realize they have not solved satisfactorily, or problems where I am not completely satisfied with the proposed solution. In either case, I'm often stimulated to try to see what I can do.

I: I think you have touched on a related question: how do you choose the statistical problems you work on?

S: I have certainly made a transition over the years in the sense that it is now rare that somebody says, "This is a beautiful mathematical problem you will be interested in," and I respond favorably. I'm much more inclined now to respond to the scientific description of a problem that I can see, or somebody will tell me if I don't see immediately, is related to a statistical problem that I might take an interest in. That was probably not the case when I was much younger. Anything that was related to what I was doing mathematically would automatically interest me. Occasionally, I still work on problems solely because of their mathematical fascination, but much less so.

I: Has it happened that after attending a lecture or seminar a problem occurred to you and you wanted to solve it?

S: Yes. I don't think very quickly. I'd say, probably two out of three times when I come out of a seminar thinking that I have something to contribute to a problem that was discussed, it turns out I was wrong. Occasionally, that can be a useful stimulus to further research. In many other cases, a seminar does not provide a problem that I work on immediately, but gets stored in the back of my mind in case a related idea turns out to be useful. In the world of mathematics, people often admit they never understood somebody else's idea until they rediscover it for themselves. I think this is a real phenomenon. You listen to a seminar or hear a series of lectures on a subject without really internalizing it until a few years later when you circle back to this area by who knows what route, view it your own way, reconstruct what somebody was trying to tell you years earlier and for a while even think it's your own idea. Eventually, you recognize that somebody else was there first. Maybe you can still make a contribution, or maybe you can't. Of course, one always hopes that one recognizes the situation before trying to publish a paper as one's own idea that was really something learned at a lecture a few years earlier.

I: Do you do direct consultation work?

S: I do a bit, but not much. There are a few people I work with who know the kind of problems I'm interested in and will be good enough not to come to me for routine assistance, but will come with a problem that interests me. This applies particularly to my colleagues at Stanford. Perhaps this is one of the main benefits of having moved there. A fairly large number of my colleagues in the Statistics Department are involved in many different problems throughout university, and they are kind enough to use me as a secondary consultant by suggesting problems that they know I would be interested in. If the problem originated outside the department, then I will often go directly to the source. This is exciting because the problems are often important, and it's much better for me than working as a real consultant for a living. Then you have to take problems for which there is a flow of income regardless of whether they are interesting or not.

I: Do you get people calling you up to ask whether you could solve this problem for them?

S: Occasionally, but not usually. I have been department chair from time to time and then it happens, not because the person knows anything about me or my interest, but simply because he finds my name somewhere in the directory or on the internet. Then I'm the first layer of contact and I play the role of trying to suggest colleagues who would be most suitable and most inclined to work on the problem insofar as I understand it. That has its own rewards but is quite different.

I: You were Associate Dean of Stanford's School of Humanities and Sciences from 1993 to 1996. What is your most memorable experience in that capacity?

S: I would say that the overall experience was quite memorable, but no single event. My role was to serve as an intermediary between the Dean of the School of Humanities and Sciences (which involves about 30 departments: humanities, social sciences, natural sciences) and the six natural science departments. The reward to me was to learn what was going on in the science departments. Part of the job that I did not particularly like was learning the enormous cost of doing modern laboratory science. I'm very thankful that I am not a laboratory person although I can also see the excitement of doing laboratory work, being closer to the scientific problems than a statistician can be, even for one doing genuine applied statistics. Lab scientists generate lots of data, and without them there wouldn't be any statistical data analysis. But modern science is an enormously expensive business and part of the job of the dean's office is to help allocate resources. You never can make people happy when you are allocating scarce resources. Learning why scientists want the resources and trying to prioritize competing requests is interesting and stimulating. It was fun trying to figure out what different people were doing, where the quality lay, what should be supported or what not. But you can never provide all the resources you want to, and you never learn as much about what is going on as you want to. You sometimes think that if you spend a few more hours, you would really make a better decision. But in the end you are forced by schedules and so forth to make decisions even when you don't understand things completely, and then you can make people upset. There are ups and downs. I'm happy to be back in my role as a scientist, which I find much more interesting.

I: Do you think statisticians are indispensable?

S: I think they are very fortunate to have the opportunities to play as many roles as they do. They are dispensable if they abdicate their responsibilities to participate in the general scientific enterprise to the extent that scientists find it easier and more satisfactory to do their own statistical analysis. But it probably also works in favor of statisticians that they are very inexpensive. It may not make sense for first rate biomedical scientists to devote a substantial part of their time to thinking about statistics if there are helpful statisticians available. You can have a first grade mathematics and statistics departments with much smaller investment than a first-rate chemistry department.

I: The humanities and the sciences are under the same school at Stanford, but they seem to be incompatible.

S: There is a constant argument as to whether they should be broken up. In the United States, the Stanford arrangement is not unusual, but it is also not universal. One somewhat interesting feature of being an associate dean was to learn about different administrative structures in different universities, and which problems the structures help to solve and which ones they don't solve. For example, I was on a review committee once for the Department of Statistics at the University of Chicago. At that time I was just beginning and spent some time talking to the long-term dean of the School of Physical Sciences at Chicago, which has a quite different structure from our School of Humanities and Sciences. For example, their School of Biological Sciences includes a medical faculty. At Stanford there are several "biology" departments, one in the School of Humanities and Sciences and several in the medical school. You would think that certain problems that arise at Stanford might have been solved by the different structure at Chicago. But it seems that while some problems are alleviated others are created, and still others exist with either administrative structure.

I: Were you able to bridge the gap between the scientists and the people in the humanities?

S: For most of my time in the dean's office, my main concentration was on the science departments. I didn't put in as much effort interacting with the humanities departments. For a short time I was put in charge of the Philosophy Department and the interdisciplinary program on ethics in society. I have occasionally thought that I am a "closet" philospher but fortunate that I don't have to earn my living that way, so I don't have to be rational, or consistent or possess other qualities we expect of philosophers. This was a very interesting experience even though I found it difficult to make informed judgments and came to rely a great deal on telephone conversations with faculty at other universities.

Theodore and W. Hugh

Logic and Mathematics

Theodore Slaman and W. Hugh Woodin have recently made important contributions to logic, especially to recursion theory and set theory respectively.

Slaman did a bachelor's degree in physics at Pennsylvania State University before going to Cambridge, Massachusetts to do his doctorate in mathematics at Harvard University. He taught at University of Chicago from 1983 to 1996 and subsequently at University of California at Berkeley, where he is a Professor of Mathematics. He has received the Presidential Young Investigator Award and the Alexander von Humboldt Research Award. He has been invited to give lectures at major mathematical meetings such as the International Congress of Mathematicians, meetings of the American Mathematical Society, British Mathematics Colloquium and the Association of Symbolic Logic's Logic Colloquia. He has given the Gödel Lecture and has been invited to give lectures and engage in research collaboration

Slaman

Woodin

in many parts of the world, in particular Japan, United Kingdom, Germany, France, Canada, China, and Singapore. Slaman is best known for his contributions to recursion theory.

Woodin obtained his bachelor's degree in mathematics from California Institute of Technology and doctorate from University of California at Berkeley. He taught at CalTech and then at University of California, Berkeley, where he is Professor of Mathematics. He has received many awards and grants for his research work, among them the NSF Principal Investigator, SERC Senior Visiting Fellowship Research Grant, Presidential Young Investigator Award, Sloan Research Fellow, Carol Karp Prize, Humboldt Research Award, Miller Research Professorship. He is a member of the American Academy of Arts and Sciences. He has been invited to give lectures at the International Congress of Mathematicians and in many parts of the world, in particular, UK, France, Netherlands, Italy, Spain, Canada, New Zealand, Germany and China. He is well-known for fundamental contributions to modern set theory and large cardinals.

Slaman and Woodin were interviewed at the Department of Mathematics on 12 June 2004 when they were invited to participate in the Institute's workshop on *Mathematical Logic and its Applications* (17–18 June 2004). The following is an enhanced version of the edited transcript of the interview in which they share their passion for and fascination with the highest form of abstraction in one of the possibly most remote area of knowledge at the boundaries of mathematics, logic and philosophy.

Imprints: Why did you choose logic and set theory in your graduate studies?

Theodore Slaman: When I was an undergraduate, I majored in physics and I took courses in physics and pure mathematics. But nothing fit my way of thinking as well as mathematical logic. I cannot say how logic felt about me, but I was attracted to it from the beginning. It was an intuitive rather than rational decision.

W. Hugh Woodin: When I was an undergraduate, I began in analysis and the problems in analysis that I was interested in led naturally to set theory. So I studied set theory.

I: Mathematical logic and set theory seem to be linked together like inseparable Siamese twins. Why is that so?

W: Methods from logic form an essential part of the study of set theory because the study of definable sets is such an important part of the subject. In that sense, logic is very much a part of set theory.

S: There are many natural questions in set theory that cannot be settled within a naive mathematical setting. The best example is probably the continuum hypothesis. Gödel and Cohen made a big breakthrough by showing that the continuum hypothesis is neither provable nor refutable within the axioms of Zermelo–Frankel set theory. It is a mathematical question in set theory but you have to use logic to understand it.

I: Logic seems to be a closed book to many people, even to many mathematicians. What do you think is the reason for this?

S: In fact, much of mathematical logic percolated into the general mathematical consciousness without the general mathematician's being aware of it. For example, logic is instrumental in the analysis of solvability problems for Diophantine equations. Given a polynomial with integer coefficients, can you tell whether it has integer solutions? A lot of people contributed and then Matiyasevich provided the final step in the 1970s to show that the answer is "no": there is no such algorithm. Of course, before Matiyasevich's theorem can be proven, one has to analyze what it means to have an algorithm. That preliminary work was done by Gödel, Church, Turing, and others in the 1930s. The very recent result that primality can be decided in polynomial time is something any contemporary mathematician should be able to understand. Again, one is building upon what it means to be computable, what it means to have an algorithm, the model of computation using Turing machines. All of that comes from logic. These days, there's a concrete feet-on-the-ground understanding of logical issues.

I: Probably for the new generation of mathematicians, but for the older generation of mathematicians, they are quite happy that they never had a course in logic.

S: I recall taking a variety of courses in analysis, topology, algebra, and geometry, and I am very happy about that. Yes, there are mathematicians who are happy that they don't know some branch of mathematics. But with a different attitude, they would have a richer mathematical experience. People should be happy for all the courses that they have taken.

W: I think the Gödel incompleteness theorem is one of the greatest theorems of the 20th century. How could one be happy about being ignorant of it?

I: They may know it but only at the superficial level. Even now, a first course in logic is not part of the essential mathematics curriculum, even in the second year.

S: It's true that in most universities a course in logic is not required but usually it is available.

I: Do you think that logic itself could be subject to immutable logical laws? How do we know that the logical processes that we use in mathematics follow the genuine laws and not some approximation of them?

W: I think that if one comes up with a proof that is a formal proof, there is no question that you have proved what you have set out to prove. In that sense it is a sufficient condition. Of course, no one exhibits a formal proof. On the other hand, I do happen to believe that there are truths of number theory or set theory which are not formal consequences of the axioms. But it's not that we are missing a logic to discover those truths. We discover those truths by understanding formal consequences of the axioms and by following our intuition. But, I don't think there is a super logic that transcends classical logic waiting to be discovered and things you can prove in super logic from the axioms give you more truth than the formal truth.

I: If I remember correctly, when a person defines "implication", it is not a "real" definition, it's just a symbol and somehow it assumes that the person already knows what it means. It is one of those things that you don't define. It seems to me that the logical process is taken for granted right from the beginning.

S: Sure, a logical process is taken for granted at the beginning. It's similar to taking the integers for granted when introducing the axioms for a ring. The point is that we analyze logic with the same mathematical precision and the same success that we achieve with the integers.

W: The issue is if you want to show that a problem is unsolvable, to make that precise you have to formalize reason. Otherwise, how do you get mathematical content into the statement like "the continuum hypothesis cannot be solved from the axioms of set theory"? You have to set up some formal system of logic and there are many different ways to do it.

I: Is there a unique system of logic?

W: No, but I think whatever approach you take, you are going to end up with the same collection of unsolvable propositions.

I: Has logic or set theory been able to solve any long-standing problems in areas like analysis or algebra that are not essentially of a logical nature?

S: One way to give a logician's solution to a problem would be to show, to say, that the problem does not have a solution within the axioms of set theory — that the basic principles used by people in the field who originally formulated the problem are insufficient to settle the problem — in the same way that the axioms of a group are insufficient to settle whether the operation commutes. There are lots of examples like that. Another way that logic could be used to settle a problem would be to actually give an answer, in the usual sense, which rests upon perspectives or techniques which are intrinsically logical, having something to do with language, definability, and so on. In the past few years, Hrushovski has brilliantly applied model theory to algebra and number theory. Of course, ideas flow in both directions; Sela recently solved a long-standing problem in logic by applying ideas from topology.

W: I give you another example. When I was an undergraduate, I became interested in set theory and was given a summer research project on a well-known problem of Kaplansky in Banach algebras to think about. I did an analysis which led to the theorem of Solovay that it was unsolvable. It was a non-logical problem and the answer is impossible within set theory.

I: What exactly is that problem?

W: The simplest formulation is this question. Consider the set of continuous functions on the unit interval. This is naturally a Banach algebra with the norm being the sup norm. Suppose you have an algebra homomorphism of that Banach algebra into another Banach algebra. Must it be continuous? You can recast the problem as the following. C[0,1] is a Banach algebra under the sup norm. Suppose you put another norm on it that makes it a normed Banach algebra. Must the norm topology be the same? In other words, are all algebraic norms on C[0,1] topologically equivalent? This problem is independent. If the continuum hypothesis is true, then there exist mutually inequivalent algebraic norms on C[0,1]. However, it is possible to build a universe of sets so that there is only one norm up to topological equivalence.

In fact, there is an area where methods of logic and set theory have led to a solution of a classical problem in analysis and not just by showing that the problem is unsolvable. Here there are various ways to cast the problem. A nice way of doing it is to use projective sets which are the sets of real numbers which can be generated from the Borel sets by closing under continuous images and complements. These were studied extensively in the beginning of 1900. By 1925, I would say, two kinds of questions had emerged. One was a measure-theoretic question. Are the projective sets Lebesgue measurable? The second was a structural question called the uniformization question: given a projective subset of a plane, is there a function whose graph is contained in that projective set and whose domain is the projection of that projective set? These are the two classical questions about projective sets. It turns out that both questions are unsolvable in set theory. We can say now that we have solved the questions. They are unsolvable in set theory because the axioms of set theory are insufficient.

The study of the projective sets is really second-order number theory. Beyond the integers, the next structure you might want to look at is the set of integers together with all sets of integers. Many fundamental questions you can ask there are unsolvable, even in set theory. The measure problem for the projective sets, level by level, is a question of second-order number theory. So the question is really what are the axioms for second-order number theory? We now know what they are, and it took many years to come to that understanding. In some sense, we have found the axioms for second-order number theory that correspond to the Peano axioms for number theory. So here we have an example where classical problems have been solved by other than purely formal means.

I: It would be dramatic if one could use set theory to settle the Riemann Hypothesis. Any chance of doing that?

W: Well, the Riemann Hypothesis is formally equivalent to the consistency of a theory. So it could be equivalent to the consistency of the axioms of set theory. If that is the case, probably set theory will be used. But there is no evidence that the problem cannot be solved just on the basis of the axioms for number theory.

There are examples of algebraic questions that have been solved using set theory. There is a very nice example dealing with free left-distributive algebras with one operation. The algebra generated by one element is left-distributive: $a*(b*c) = (a*b)*(a*c)$. There are some natural questions you can ask about this. It turns out that the natural model for this comes from set theory and that model was used to gain insights into that algebra. In particular, the first solution to the word problem came from this natural model. Subsequently, it was done without it. But there are still questions about this algebra that have to do with whether the free algebra is an inverse limit of a canonical sequence of finite approximations. The only proof known still uses the natural model from set theory, but there is no evidence that it really needs it. This is an example where a truth was discovered first by invoking very powerful axioms. And there are other examples, such as those coming from the determinacy of infinite games.

S: Determinacy talks about two players and a given set of reals. You use the set to organize a game between the players. The players alternate playing integers to form an infinite sequence. If the sequence is in the set, the first player wins. If not, the second player wins.

W: Suppose the given set is a subset of the unit interval and we are going to play integers between 0 and 9. We create the decimal expansion of a real from the integers played. Player 1 wins if that real is in the set, otherwise Player 2 wins.

S: Once we have defined the games we can ask whether they are determined, i.e. if one of the two players has a winning strategy. For example, if we start with the set of all reals, then the first player is going to win. Any strategy will work. The theorem that any Borel set is determined has a remarkable history and is due to Tony Martin. The first proof used measurable cardinals (so, very strong principles from set theory). Later on, Martin proved the determinacy of all Borel games using just the axioms of set theory.

I: What do you think is the greatest contribution of logic to computer science?

S: My answer to that would be the definition of a computable function. It grew purely from logical considerations. Turing presented his model of a computing machine as part of his argument that if a function could be computed at all, then it could be computed by a machine of his sort, namely a Turing machine.

I: This was introduced before the first computers...

S: Yes, the work on the foundations of computability took place in the 1930s and predates actual computers considerably. There are other examples. Logic has to do with the analysis of language and definability, the resources of definability and algorithms. If you look at logic as having a scope that wide, then computer science is looking at a certain section of logic. So you cannot in any way think of computer science without logic. Theoretical computer scientists are analyzing different levels of complexity of computation, and that's all logic.

I: We know from Gödel's incompleteness theorem that there are true results about numbers that cannot be proved within arithmetic. Is it possible to produce concrete, non-metamathematical statements of this nature?

W: A famous example concerns Goodstein sequences. Goodstein published a paper in the forties presenting a number-theoretic fact which he thought was a good candidate for statements which cannot be proved from the axioms of number theory. In the '70s, that was shown to be the case. You start with a seed number *N* and generate sequences of integers, called Goodstein sequences. The theorem is that the sequence is eventually 1. You start with your seed number *N* and expand it to base 2 as a sum of powers of 2. Then you take every exponent and expand it as a sum of powers of 2 — exponents of exponents and so on. Now change all the 2s to 3s: this defines a new number expanded in base 3. Subtract 1 and re-expand to base 3. Now replace the 3s by 4s. Subtract 1 and re-expand to base 4, and so on. The theorem is that you will eventually reach 1. It may take a long time but you will always reach 1. If you use your calculator and start with 33, you will overflow your calculator.

S: Whether the sequences converge is not a metamathematical question. In theory, you can sit down and write them out. Remarkably, the proof that the Goodstein sequences converge involves the transfinite ordinals.

I: But practically, it is not possible to compute those numbers.

W: Well, you can do it with small numbers, like 5. Very soon it gets very complicated. This is predicted by the metamathematics.

I: Are there other examples?

W: Harvey Friedman has other examples which are quite interesting. In fact, there are lots of examples of number-theoretic statements which are not solvable within number theory, which are true and which are purely combinatorial.

I: What about the Goldbach conjecture?

W: Well, it could be equivalent to the consistency of ZF or very strong theory but there is absolutely no evidence for that.

S: Or it could be false.

I: Probability has been successfully axiomatized by measure theory, which is essentially about set theory. Yet it is not clear that this gives a true understanding of randomness. Has there been any progress on shedding some light on the intrinsic concept of randomness from the logical point of view?

S: I think there has been a lot of progress. You can trace it back to Kolmogorov. He had this nice idea. He looked at infinite sequences. A random sequence should be as "complicated" as possible. It should be "unpredictable", it has no historical pattern. It's just noise. But, what does it mean to have no pattern? What is a pattern? That way of speaking has to be made mathematically precise. To have no pattern means that there is no algorithm that will describe the sequence. It has a pattern if it behaves according to some algorithmic law, like the sequence of primes — you can write a program that will give you more and more of the primes. So the sequence of primes is not random. It does not have an obvious pattern such as "every other digit is 1", but it does have a pattern — a computational pattern. Now, if you want to say that a sequence has no pattern at all, then you can say there is no way to compute a pattern. Kolmogorov brought a logical perspective to the concept of randomness.

I: That has nothing to do with probability?

S: Probability and randomness seem closely related to me, but logic has had more to say about randomness. It is not hard to describe. A finite sequence is said to be "simple" if there is a program that will compute the sequence digit by digit and the length (the number of symbols) of the program is less than the length of the sequence. A random infinite sequence has the property that after some finite clustering, none of its initial segments are simple. That's the notion of Kolmogorov complexity; a sequence can be descriptively random. That definition of "random", with some technical adjustments, is equivalent to one asserting that the sequence does not belong to easily described sets of measure zero. On the one hand, you have the measure-theoretic definition of "random". On the other hand, you have the property about an individual sequence that says that it has this bit-by-bit indescribability. The equivalence of the two is very pretty.

I: What are the greatest advances in logic and set theory in the last century?

S: First, a caveat. The caveat is that the further in the past you look, the easier it is to tell the contributions that had the most impact, changed the way people think about mathematics. Those would be the greatest. I would say the identification of what it means to be provable and the Gödel completeness and incompletes theorems are great in that way. Gödel's incompleteness theorem states that the method of proof, which he showed captures logical implication, is insufficient to axiomatize the basic facts about number theory. That's a real advance and it changed the way people thought about the mathematical enterprise. A 19th century mathematician might think, "We know what the proofs are and we should be able to find the correct set of axioms on which to base all of mathematics." This mindset was completely gone by the middle, certainly by the end, of the 20th century. Giving a clear definition of "computable function" was another great achievement. With that available, we can prove theorems about algorithmic solvability and unsolvability within a mathematical setting.

W: In set theory, I think Gödel's isolation or identification of the constructible universe is certainly one of the greatest achievements in set theory in the 20th century. Cohen's discovery of forcing was also one of the great achievements in set theory of the last century. I think the validation and understanding of projective determinacy is also an important advance, but even among set theorists it is debated whether the axiom should be regarded as true.

S: Professor Woodin would have to be modest about this since he was involved in that work. I do think logicians looking back 50 years from now would see it as something great.

I: There were many scientists and mathematicians who started off as logicians, like Norbert Wiener, John von Neumann.

S: And Saunders MacLane.

I: Any advice for graduate students in mathematics?

S: There are two things you have to do when you are a graduate student. First, read papers (a lot of them), work out the details, learn the body of mathematics, and learn as much as possible of applied mathematics, physics and biology. Then, identify a field of interest and learn it in depth. It's very important to choose the questions to study. You should have the big questions in mind. The smaller questions you work on should have a bearing on the big ones. Secondly, and maybe this should have been primarily, you have to choose the right advisor.

I: What are your hobbies when you are not doing mathematics?

S: The richest part of my life is the company of my family. Mathematics is second to that. There is no candidate for third place. Third place is always going up for rent: running marathons, fixing my house, programming my computer, doing administrative work at the university.

W: Sleeping.

Terry

Good Gene Hunting

Terry Speed is world-renowned for his important and numerous contributions to the applications of statistics to genetics and molecular biology, and in particular, to biomolecular sequence analysis, the mapping of genes in experimental crosses and human pedigrees, and the analysis of gene expression data. A member of the NIH Genome Study Section from 1995 to 1998, he investigated fundamental problems arising from the Human Genome Project.

He has received numerous honors from the world's leading scientific bodies and has been invited to give lectures on his research; in particular, he was a Wald Lecturer at the US-based Institute of Mathematical Statistics. He has been on the editorial boards of international statistical journals, and currently of the *Journal of Computational Biology*. He was President of the Institute of Mathematical Statistics.

Speed

He holds joint positions in the Department of Statistics at the University of California at Berkeley and in the Division of Genetics and Bioinformatics at the Walter and Eliza Hall Institute (WEHI) of Medical Research in Melbourne. Each year, he divides his time equally between the two organizations.

The following is the result of an interview of Terry Speed conducted in three stages: an "electronic interview" shortly before he came to the Institute in January 2004 as invited workshop lecturer of the Institute's program on *Statistical Methods in Microarray Analysis*, a face-to-face meeting at the Institute and a final "electronic interview" after he returned to Berkeley. The result is a frank and insightful revelation of an intellectual journey from a statistical beginning shrouded in abstract algebra and mundane experimental designs to one of the world's principal centers of activity responsible for the unfolding of one of the most dramatic scientific dramas of the 20th century.

Imprints: What did you do for your PhD?

Terry Speed: My thesis was in algebra, being entitled something like *Topics in Distributive Lattices*. I'd started thinking I might be able to do something interesting on (elementary) probability theory within a non-classical logic known as intuitionism, and ended up studying aspects of the algebraic structure underlying that logic. I hadn't done much algebra in my undergraduate degree, so I was to some extent just catching up. It was fun. I found that I liked algebra, which was new to me, and I have continued to enjoy the algebraic aspects of what I do more than the analytical ones.

I: When and how did you get interested in applying mathematics to biology?

S: I was always interested in mathematics and in biology. My aim on leaving high school was to go do combined science and medical degrees, and to go to work in medical research. This was in 1960, the year Macfarlane Burnet, then Director of my current institute, shared the Nobel Prize for medicine. On arriving at university I found that I was ok at mathematics, but less so at practical science such as lab work in biology (e.g., dissection of rats and mice, looking at cells down a microscope, etc). So I switched entirely to mathematics and statistics, but included some genetics, which didn't have labs. That was given by a former student of R. A. Fisher: Peter Parsons. I wrote an undergraduate statistics thesis on a topic of R. A. Fisher's: the survival of mutant genes in populations, elementary theoretical population genetics. Throughout my PhD I was surrounded by outstanding people in that field, but I resisted the temptation to join in. At the time I was hooked on pure mathematics... so much to learn, so little time.

I: When you first started to apply mathematics to biological problems, was there any beacon at that time in showing the way, or did you have to hack a path through virgin jungle, so to speak?

S: From my PhD until the time I went to Berkeley, I did very little on quantitative biology apart from what cropped up in statistical consulting and in collaborations. It was mostly very classical, with a couple of exceptions (e.g., some baby pedigree analysis): I doubt that DNA ever got mentioned. When I was in the maths and stats part of CSIRO I was conscious that Australia's leading "genetic engineers" were also in CSIRO, doing fancy DNA-related stuff near me in Canberra. Our division did statistical consulting for them, but there was nothing involving DNA. When I asked them could we statisticians get involved in this "gene-splicing" and perhaps help them with the quantitative aspects, I was told fairly firmly: no, there's no statistics of any kind in that research, go and design some more agricultural experiments for those old-fashioned folk over in the other building, and leave us to our high-tech stuff. So we did. Of course they were wrong... elsewhere in the world bioinformatics was being created around that time (early 1980s). Naturally, I wonder how things might have been different if they had been more receptive... (It's always good to get into a new area early on, so I tell my students: when the basic problems are still unsolved!)

All this changed when I went to Berkeley in 1987, for there the "routine" statistical consulting that came in the door involved DNA: molecular evolution, intragenic recombination, and other topics, still of interest today. Then I realized I had to catch up with about 30 years of molecular biology, and fast, if I wanted to have a chance of answering the questions they brought to our consulting service. Incidentally, this is one of the many good things about doing statistical consulting: you never know what might walk in the door, and it really can give you new interests, and change your research directions. Of course, it is scary too, because you are on unfamiliar or only vaguely familiar territory much of the time.

I: When you first went to Berkeley in 1987, what was the state of computational biology like? Did you have any hunch that something momentous was in the brewing?

S: "Momentous" is a bit strong. It was clear that very interesting things were happening on the genetics and molecular biology front in 1987. PCR had just been invented, and was helping people generate lots of interesting data, the Human Genome Project was starting to get talked about, mitochondrial Eve was in the air (later on the cover of *Time*), the first large-scale human genetic map was published, and so on. A big player in the genetic mapping world was Eric Lander, who I was told was a former pure mathematician. (He is now an even bigger player: a key member of the public Human Genome Project, now forging ahead with grand plans in this post-genome era.) I missed his visit to Berkeley, but got to read his papers nevertheless. Also, I knew that Sam Karlin was very active in the field, and I quickly became aware of the many contributions of Phil Green, another former pure mathematician, and Mike Waterman, an ex-probabilist. So people from our area were already key players, and I might have thought, "Why not me too?" But I just plugged away, trying to find a niche, thinking that perhaps I was already too late, that all the basic problems were solved! However, the forces that kept me involved were biologists. They were (and always are) so keen to use the latest and best methods, to be first to use a new technique in their particular corner of the subject, so if you are willing to try to help them, as consulting statisticians tend to be, you get swept along. You find yourself explaining and using Lander and Botstein's program for QTL mapping, Phil Green's CRIMAP, Sam Karlin's BLAST calculation, Mike Waterman's alignment algorithm, and with a little luck, you will eventually have an idea, or get a student interested, and away you go. In the decade of 1987–1997, I learned two things: that the basic problems were not always easily solved, and that you are never too late for the next train (as Piet Hein used to say). In the mid 1990s, microarrays came along, and after a while just watching, I tried my luck with some basic problems.

I: I understand that Berkeley now has a number of groups and programs that attract mathematicians, biologists and others to do interdisciplinary research. Could you tell us something about these groups and programs?

S: That is a tough question to answer briefly, and we have a web site devoted to answering it. I'd rather just refer the interested reader to it: http://computationalbiology.berkeley.edu/

Perhaps I should add that we do have a wealth of what I abbreviate as "compbio" activity at Berkeley. So much that I could spend all my time going to seminars and sitting in lab meetings, and never find time to do my own work. That's almost my fate, but I have a number of wonderful PhD students who keep me active. I get people from other departments onto their thesis committees, I'm on the committees of students from outside statistics, and I ask my students to talk to or even collaborate with students from other departments and go to their lab meeting, and I have a small number of close collaborators of my own. If I play my cards right — and I'm still learning — I can benefit from this profusion of activity, and not get completely swamped.

I: If a mature mathematician wants to work on fundamental problems in biology or computational biology, how much biology does he or she need to master?

S: Short answer: lots. Longer answer: even more. Longer answer still: how much does a person need to know about the internal combustion engine to be a good motor mechanic? How much mathematics do you need to know to teach a course on group theory? Of course you can get by, perhaps get a few papers published in journals with little or no biology, if someone else has done the job of phrasing the problem in mathematical form for you. Then you may solve it, but that's not doing computational biology, that's doing mathematics. No offence meant, but I sometimes say (and am doing so again now) that if you ask that question, you are already doomed (not to go far in computational biology). To put it another way, if you are not genuinely interested in biology, at least to the extent that you enjoy learning what you can about the area in which you are working, then it's probably not a great idea persisting in that area. Do what you like, I say.

I: Has statistical genetics discovered any general rules and principles about the mechanism of gene formation or combination? Do you see any parallel in the state of biology now and the state of physics 100 years ago in the sense that there are many empirical rules and observations but a paucity of underlying theory?

S: Genetics has lots of general rules, and lots of exceptions. As for your second question, I don't accept the implicit assumption that physics is a useful model for biology. Perhaps it's just my lack of imagination, but I don't see us understanding life any century soon. We might think that physics has made great leaps towards understanding the universe at the level of particles and the universe, what with nuclear weapons, space travel, and laser scanners at supermarkets, but in my view this is easy compared to understanding cells. Wait a few centuries and you'll see what I mean.

I: Could you give us some idea of the problems on which you are working. What is your most memorable achievement?

S: That's hard for the reasons I outlined above. I'm not working on the Speed program or conjecture or hypothesis, I'm thinking (when I get the time) what I can do at all with some problems, and what I can do a little better with others. What are those problems? Well, they are always parts of bigger problems that belong to other people: what genes, if any, have their patterns of expression changed in the brains of people with bipolar disorder, in comparison with otherwise similar healthy people. What gene expression patterns change as we age? Finding ways of distinguishing real from apparent gene expression differences, in a variety of contexts, occupies a good deal of my time. At the "continuing challenge" level, that is my aim: to distinguish the real from the apparent. This, of course, is a statistical problem with no single, final answer. There are others: I help people analyze their data to get better measurements of the things they want to measure. Then I sweat over questions such as: how can we tell this method of analyzing the data gives a better measurement than that method? My most memorable achievement? I'm still waiting. I hope to make a little progress on problems like the ones I just mentioned, and if I did, that would be memorable.

I: It is often said that this century will be the century of molecular biology. In your opinion, how much of this is hype and how much of it is scientifically justified?

S: Perhaps the best answer here is yes. That is, yes, it is hype, and yes it is scientifically justifiable. Beyond that, I don't care to go. But don't count physics out.

I: Is there some kind of mathematical definition for a gene?

S: The history of the notion of a gene is almost the same as the history of genetics, at least in the period since the field had a name, which is essentially the 20th century. So bear in mind that for biologists, the notion of gene is an ever-changing one. In the 15 years or so since the advent of large-scale genome sequencing, there has in essence been a mathematical definition of a gene, because people have used mathematical models to "find" genes — putative genes might be a better way to put it — in genomic DNA sequence. Of course, the definition of a gene computational biologists use is at best a crude approximation to what biologists understand by the term "gene", but if the computational definition does the job, i.e. if it finds "real" genes, no one is going to mind too much. Nevertheless, a model that works in one context is still just a model, and is necessarily different from the real thing, so no one should expect that the computational person's gene model embodies all important aspects of a gene.

With that preamble, let me say that a (protein-coding) gene could be defined as a DNA sequence structure consisting of a number of parts (promoter, transcription and translation start and stop signals, exons, introns, etc), each having characteristic features (base composition, dependence, specific motifs, ...) and all arranged in a characteristic way. The complete specification today might be given via a generalized hidden Markov model (GHMM), with a given state space and set of parameter values, but do realize that this would hardly be recognized by a biologist. And furthermore,

Charles

The Invariant, the Direct and the "Pretentious"

Charles Stein is considered to be one of the most original thinkers who made fundamental contributions to probability and statistics. He has received many honors and awards and is a member of the National Academy of Sciences, USA. He has given many invited lectures, notably as plenary speaker of the International Congress of Mathematicians, and as the Institute of Mathematical Statistics Wald Lecturer, Rietz Lecturer and Neyman Lecturer. He is now Emeritus Professor of Statistics at Stanford University and continues to be active in research in statistics. On 22 March 2010, a symposium in probability and statistics was held at Stanford University to celebrate Stein's 90th birthday and was jointly organized by Stanford's Department of Mathematics and NUS's IMS.

He was interviewed on 26 August 2003 at the Institute when he was the guest of honor of the Institute's program *Stein's Method and Applications: A Program in Honor of Charles Stein* (28 July–31 August 2003). The following is based on an edited transcript of the interview and subsequent follow-up by electronic mail. Here he reflects on his work and expresses his views on teaching, research and statistics.

Stein

Imprints: The first question concerns the statistical work which you did for the Air Force during the war. Presumably statistics was then not yet established as a rigorous discipline. What was it like to do statistics without the benefits of its modern foundations? Was that work instrumental in leading you to think about basic questions in statistics?

Charles Stein: First, I should say that I am strongly opposed to war and to military work. Our participation in the Second World War was necessary in the fight against fascism and, in a way, I am ashamed that I was never close to combat. However, I have opposed all wars by the United States since then and cannot imagine any circumstances that would justify war by the United States at the present time, other than very limited defensive actions.

Statistics was already a well-developed field by 1940, going back to work of Gauss, Laplace, Galton, Karl Pearson, Student, Fisher, Neyman and Pearson, and many others. On the other hand, one can argue that statistics is not yet established as a rigorous discipline. I do not think that my work on the verification of weather forecasts had much influence on my later work.

The actual type of work I did at that time was not really instrumental in leading me to think about basic questions. However, we had a very strong group of people there, Kenneth Arrow, George Forsythe and Gil Hunt among them. Certainly discussions with them helped broaden my understanding of statistics, in particular, with Gil Hunt. Gil Hunt is a very accomplished mathematician and I profited a great deal from his knowledge of group theory in particular although not as much as I could have.

I: You mentioned Gil Hunt…

S: Gil Hunt is a mathematician, Kenneth Arrow is an economist and George Forsyth became a computer scientist and numerical analyst.

I: Did you do something on groups with him (Hunt)?

S: We considered the question of whether, given a statistical problem invariant under a group of transformations, there exists an invariant procedure possessing desirable properties, such as being minimal or admissible. We showed that if the group is, in an appropriate sense, composed of groups each of which is abelian or compact, there exists a minimax procedure that is invariant under that group. Realistic counter-examples came much later. The full linear group in two or more dimensions does not satisfy this condition, and in fact the conclusion does not hold. Thus, for example, it is usually inappropriate to assume automatically that a sample covariance matrix is essentially the right estimate of the population covariance matrix. The question of admissibility also came later. Unfortunately, both Hunt and I were very slow to publish, but proofs were eventually published by other people, with full acknowledgement of course.

I: Often, work done for one's PhD thesis shapes one's future conception of the field. Is this the case for you?

S: No, this is not really true of my PhD thesis, which dealt with a topic related to Wald's sequential analysis. Kenneth Arrow had lent me Wald's first work on sequential analysis and asked me whether it provided a sequential test for Student's hypothesis having power depending only on the mean, and when I replied that it did not, he indicated that his was a serious shortcoming. A night or two later, I was officer of the day and had to prepare a forecast starting at about two in the morning. After finishing the forecast, I worked out most of the details of a two-sample test which did have power depending only on the mean, vaguely similar to earlier work of Dodge and Romig on a different problem. This became my PhD thesis. Neyman was impressed, because George Dantzig, who was a student of his, had proved rigorously the intuitively obvious fact that no single-sample test can accomplish this. However, I think of this work as relatively unimportant, and it did not have much effect on my later work.

However, Wald did have a strong effect on my work. My work with Hunt grew out of Hunt's remark that, in my attempt at generalizing a result of Wald, I was essentially doing group theory, and all of my work on statistics has been in the framework of statistical decision theory, developed by Wald following ideas of Neyman and Pearson, and also von Neumann and Morgenstern. Wald also encouraged me to work on mathematical statistics while I was in the Army Air Force, and to come to Columbia after the war.

I: If you were asked to list your three most important contributions to statistics and mathematics, what would you list?

S: Certainly the work that is most important is what is called Stein's Method, which was developed further by Louis Chen, whose work inspired much other related work. The second most important is my work on invariant problems, which started with my discussions with Hunt, and continued with my paper in the Third Berkeley Symposium on the inadmissibility of the usual estimate of the mean of a multivariate normal distribution in three or more dimensions, and the paper with James in the Fourth Berkeley Symposium, which studied a reasonably good estimate for this problem. Efron and Morris also proposed an important improvement on this estimate. Others, including many of our students, such as Eaton and Loh, did important work involving unknown covariance matrices.

Most people seem to think that my other paper in the Third Berkeley Symposium is my third most important contribution. In response to the problem of estimating the median of an unknown symmetric distribution, Vernon Johns and others proposed a sensible solution, an "adaptive" symmetric average of order statistics. In response to the same problem, I wrote a rather pretentious paper in that symposium, which tried to develop a general theory applicable to this problem. I got nearly everything wrong but the paper is believed to have had considerable influence on the development of semiparametric statistical methods.

I: I think you are modest about that. You mentioned Stein's Method as your most important contribution. Many people would like to know what led you into formulating the method that is now known as Stein's Method?

S: Well, Persi Diaconis has already touched on that in his lecture. Briefly, it is that I was teaching a course on non-parametric statistics and I decided to prove what is called the combinatorial Central Limit Theorem. I could have presented the published work of Wald and Wolfowitz and Hoeffding but instead I decided to try my own approach. That involved the idea of exchangeable pairs to approximate the characteristic function. After a while, I realized that there was nothing special about the complex exponentials. I introduced an arbitrary function and thereby avoided the need to invert the characteristic function.

I: You were looking at it in a different way. Most people would just teach it in the standard way from the books or papers. Not many people would actually try to find a novel way of looking at it. It is quite well-known that teaching does contribute to research, isn't it?

S: Many people, including notably David Blackwell, have mentioned that teaching is the most important stimulant to their research.

I: Did you foresee the wide applications that your Method brought to other fields some 15 years after it was introduced?

S: No, but I guess that I always thought it ought to have applications. I never really pursued it.

I: I understand that you do not have a personal pressure to publish. Don't you think it would be a loss to the community if you don't publish the results you have?

S: I have always had great difficulty writing things up and also difficulty in forcing myself to submit something even after it is written. I suppose this has slowed progress somewhat in a few cases, and I regret it.

I: Do you have many graduate students?

S: No, I had about 10 graduate students personally. Louis Chen and Wei-Liem Loh were among them.

I: Your students will invariably pick up your ideas and extend them. In some sense, they are doing what you are doing.

S: Yes.

I: In research, which is more important: conceptual foundations or technical perfection?

S: That is a hard question to answer. I find it hard to get all the details right, and yet that is important. But I am stronger on the conceptual aspects than on the technical aspects.

I: Would you say that your approach to a problem is intuitive?

S: To some extent. I would look at problems which are not always very well clarified at first, and then I go on to clarify them, I hope.

I: What is your view on breadth versus depth in research? For example, some people work deeply on a small topic while some people have a wider interest.

S: I must say that I have not solved a wide range of problems though the problems I studied are formulated over a broad range, like the invariance problems and the question of using direct elementary methods rather than complex variable methods in probability theory. Of course this was not new, going back at least to Lindeberg in the modern period, but elementary methods had become unfashionable and my approach is much more widely applicable.

I: The next question is about the explosion in information and knowledge that we are faced with nowadays. Is it necessary to keep up with this explosion in information and knowledge?

S: For young people, it is important and, to some extent, it may be possible for them. At my age, it is not important. I am incapable of keeping up with, for example, the Fermat conjecture. I have tried reading books on it but have not made much progress in understanding it.

Everyone is exposed to some aspects of computational theory and practice, but that is more important for young people.

I: Do you use the computer in your work?

S: I have done some computing although my colleagues tend to discourage me. In a recent work on simulation, my co-authors did most of the computing.

I: I suppose you give them the ideas and they do the computing.

S: They may be stronger in computing then I am but they are also strong theoretically.

I: Could you give us your projections for statistics in this new century?

S: Not really. Clearly, the field continues to develop. The computational aspects will be perhaps even more important than they are today. There will probably be a mixture of good work and bad work. There will be a lot of statistical packages to enable people to solve problems, often without understanding them. But I am unable to anticipate the directions of the important changes.

I: What do you think of the importance of statistics in computational biology?

S: Statistics is, of course, important and people are interested in applying it.

I: In some sense, what you have done plays an important role in computational biology.

S: I have not followed the field of computational biology enough to know whether my ideas have really been useful.

I: You mentioned the computational aspects of statistics. So we are going to be faced with a lot of data generated by computational statistics and then the theory has to keep up with it. We seem to be like in physics where there is more information than theory. Do you think there will be a revolution in statistical theory?

S: I cannot anticipate that. There is one big question on which I am not really competent, and that is the extent to which elaborate models are useful in applied statistics. Elaborate models can give an impression of providing results that are in fact not justified whereas with simpler models, it is more apparent if the results are not justified. The elaborate models are like black boxes which are supposed to give you the answers. And you do not know if anything goes wrong.

I: Do you find it surprising that statistics can do so many things and solve many problems in real life?

S: No, it is not clear that it does. It certainly plays an important role but one should not put too much confidence in this claim.

I: But statistics takes the guess work out of solving problems. In the old days, you did not know what is going on and you did it by trial and error. Now statistics gives you a way of doing things.

S: It gives you a way of thinking about things, but you may not come out with correct conclusions.

Gilbert

The Changing Face of Applied Mathematics

Gilbert Strang is a prominent scholar in applied mathematics and an active promoter of mathematics and mathematical education in the United States. He has made numerous contributions to numerical analysis, wavelets and signal processing. He is also well-known for his textbooks on linear algebra and applied mathematics at the undergraduate and advanced levels. He is editor of many well-known journals and has given invited lectures throughout the world. He has received numerous honors and awards and is a Fellow of the American Academy of Arts and Sciences. He has served on many committees, in particular, as President of the Society of Industrial and Applied Mathematics (SIAM) in 1999 and 2000. He was Chair of the US National Committee on Mathematics for 2003–2004. He has been Professor of Mathematics at MIT since 1970.

He has been closely associated with NUS, having served as a member of the University's International Advisory Panel from 1998 to 2001 during a period of reorganization of NUS. He was interviewed on 18 July 2003 at the Institute for Mathematical Sciences when he visited NUS

Strang

as a guest of the Institute and the Singapore MIT Alliance from 15 to 19 July 2003. The following is an edited transcript of the interview in which he spoke about e-learning and teaching, applied mathematics in the service of society and the changing landscape of applied mathematics.

Imprints: MIT recently launched an ambitious long term project to make its courses freely available on the web. How has this project influenced the teaching of mathematics at MIT?

Gilbert Strang: Generally, the MIT OpenCourseWare is an e-learning project. I don't think anybody knows exactly how the course materials on the web are going to be used. It was by chance that I had videos taken of my linear algebra lectures (I teach a lot of linear algebra courses). About two years ago, they were video taping the freshman physics course when I came into the same room for the next hour. So, I said, "Maybe you can just keep the cameraman in the room for another hour and we will have a video tape of the linear algebra lecture." They got some financial support for it and they did it. As a result, my linear algebra course is one of the few that have videos on the web. It's absurd that I am now a movie star on OpenCourseWare, which is available at http://ocw.mit.edu. That course was in the first group and now 500 courses are available. I have two others, an applied mathematics course and a wavelets course. The linear algebra course was an early one.

From early feedback on the number of hits around the world, Hong Kong was leading in the number of users. It may be different now. I think students like to see videos which are more alive than course notes. Your question asks about e-learning and mathematics. E-learning has to be alive! The lectures have to somehow involve the student. The real challenge is for students to be a part of the learning process and not just passive viewers.

I: Is there any significant influence on people's teaching?

S: It makes teaching interesting in a new way. After my turn, they took videos of really good lectures at the Mathematics Department. For example, Professor Arthur Mattuck had given lectures for years on calculus and differential equations. I was happy that he was video taped. He is always very well organized. We are also trying to see if there is a way for us to have students answer simple questions in the class, either by pressing a button or in some other ways. You want them to follow the lecture closely, instead of sitting there till the end and then leaving. The key is active learning.

I: You were President of the Society for Industrial and Applied Mathematics (SIAM) some time ago. Could you share with us some of your experiences, challenges and achievements during your presidency?

S: I was President for two years in 1999 and 2000. For a year before, I was the President-Elect, and Past President the year after, so it was a four-year commitment. Quite a lot of time, but I enjoyed it. Applied Mathematics in the US has been upgrading its efforts in connecting with Washington. I think Singapore is amazing because you are always well connected to the needs of society and to the goals of the ministry. In the US, we were far from Washington in the past (well, still mostly so). Mathematicians are thinking more about their own work than about the goals of the Science Minister.

But now, there is more and more input to the National Science Foundation, to the Office of Management and Budget, to the Senate, to the House. This is a crucial step towards addressing the question of what applied mathematics can contribute to society. It is a challenge, and a happy experience, to speak to the House, to the Senate and to their staffs about mathematics.

One of my special experiences was getting two new activity groups started: one in the mathematics of Life Sciences, and one in Imaging Science. In SIAM, the activity groups have concentrated conferences every two years. It is really important for applied and pure mathematics to recognize what are the new directions and to help the new areas. Some problems grow more and some less, and the rate of changes varies.

The action moves into new areas like the life sciences. It's really important for SIAM to be a part of those areas. It does a lot of conference organising, but the job of the President and the Council of the society is to see what new actions and what new work the society should be doing. That depends on where applied mathematics is moving.

The society needs to establish a better image of mathematics. The engineers are always there to sell their ideas, and the physicists and biologists are also there. You have to demonstrate the ability of mathematics to contribute and you have to show that mathematicians are needed. Of course, that has to be proved by what actually happens so that the engineers or the biologists can see the contribution from mathematics.

I: The term "applied mathematics" seems to change over time and place. What does it mean nowadays?

S: There are different parts of applied mathematics. One person is probably not naturally prepared to work in all these parts. An important part is problem modeling. Given the physical problem, produce differential equations that capture the essential facts of the physical problem. Modeling is a big step and I am always impressed by people who do that well. Modeling may not be rigorous but people who do it will agree when it is done well. There is definitely a standard of good and bad in modeling, but it is not a theorem-proof standard. There is an intuitive agreement that, yes, the important parts of the problem are correctly accounted for.

I: How much rigor do we need to maintain in modeling?

S: Maybe not rigor in a sense of proving, but it has to be convincing. The expert in the field might say, "Ah-ha, you missed this important aspect of the problem" and that will affect the solution. There is a standard but it's not exclusively logical. After the model is created comes computing. "Computational science" is now the word that includes both these aspects. You understand the area of application and you can carry through the computing to obtain and understand the answer. But often that is shared. One person understands the application area and the other person says, "Just give me the equation and I will work on the solution."

I: Does that mean the applied mathematician must have two modes of thought?

S: Yes, if he or she is complete. The modeling mode and, we could say, the computing mode.

I: Not many people are able to do that.

S: Unfortunately so but it doesn't have to be. It's good if it is all in one person because then that person sees the full picture. But very often, say in linear algebra, there are many people who will ask, "I have a very large symmetric matrix. How do I find the five smallest eigenvalues?" To the expert in numerical algebra, this is a question independent of where the matrix comes from. There is a further stage of creating good software that could be permanent and that other people could use. So, there you really need to understand computer science. So, we have the modeling stage, the solution stage and the software stage. The third stage seems far from classical applied mathematics, but it is part of the whole process.

I: So it seems that the computer is really indispensable?

S: It is now, yes. In the past, people could get amazing information about the solution by analytical methods and by perturbation methods. That was done by the mandarins of applied mathematics. Now it's more democratic and it's a larger world. Engineers are now part of the modeling world, using algorithms and sometimes creating algorithms. For example, the finite element method is a famous example of an idea that was already in the mathematics papers by Courant and Feng Kang but was really brought to importance by the engineers.

I had an interesting conversation with Nick Trefethen in Oxford on whether mathematicians have made the key contribution in scientific computing. He convinced me that it was probably so. He went down the list of main algorithms that are really dominant. One example would be the Fast Fourier Transform that is used almost everywhere. It came from, well, maybe statisticians (in John Tukey's case) and from mathematicians (Gauss was the first). There is a whole range of crucial algorithms in the 20th century like the fast multipole methods, stability problems for finite differences, and applied mathematicians have been at the heart of the progress of those methods.

I: Who provided the key ideas?

S: I have never before made the list that Trefethen has: what are the key ideas and who brought them to birth? I like engineering activities and certainly engineers deserve credit. Wavelets is a good example where work in applied mathematics has changed the direction of some area of pure mathematics. Wavelets grew out of signal processing, which is a big user of mathematics. Wavelets have their own unique direction and depth in the pure mathematics sense. Progress in that area continues. That is also well reflected in the Institute's programs and in the wavelet center here.

I: It seems that applied mathematicians would need to work with somebody else.

S: Often, yes. The new problems and excitement are now in the life sciences and biology. You need the biologists, of course. They know what the problems are but they need the mathematicians too. The border between pure and applied mathematics becomes quite permeable. It is very satisfying to contribute to the solutions that other people want. That really is a good feeling. And pure mathematicians often see those patterns and relationships that are fundamental to mathematics.

I: Is it necessary to give special training in order to make somebody into a good applied mathematician?

S: Well, I don't know. I think that the modeling ability is partly genetic and in-born. The computing skill comes partly from experience — actually doing the computing, getting the answers and improving the algorithms. It is a field where the supply is far behind the demand. The demand is everywhere and I would like to see more and more students having the fun of contributing to a team of scientists.

I: Do you think that the ability to model is unique and that not everybody can do it?

S: Unique, yes. But it must be possible to teach something to those of us who have not got it. I suppose I may be feeling about modeling the way so many outsiders feel about mathematics: that it is a mystery. How could people do this? But mathematicians don't see it as a mystery. They see it as a natural ability. Probably modeling should be just another ability that could be developed. But the very best modelers seem to be natural and not taught.

I: Do you consider modeling to be more of an art?

S: Yes it is. The computing part (the algorithmic part as well) takes on its own creativity. It is an area where you need to experiment and if you have an algorithm, you prove that it is good by using it. A paper that just suggests an algorithm that is never tried is not really acceptable. The proof is in the solution.

I: What are the big problems in applied mathematics?

S: One of the very big areas has been computational fluid dynamics, CFD, and in the extreme case, turbulence, where the parameters are approaching the departure from a smooth flow. That is a big problem and there is good progress. There will always be wonderful problems. But I may not be the right person to name all the big problems.

I: Is turbulence a computational problem?

S: Well, you could say that it is partly computational. You need to invent new methods. There is another big area currently: multi-scale computation. Typically, you create a mesh. You have a step-size close to the time scale or the length scale of the problem. But if the problem has a length scale that extends from the size of an atom or molecule to the size of the body or the size of the earth, you cannot use the molecular length scale. How do you make the computation at the molecular length scale acceptable for a model of the heart? Somehow you have to find a smart way to see the effects of very, very small scale events at the macro scale.

I: What about artificial intelligence? It used to be a hot topic.

S: Yes, it did. Maybe it did count as applied mathematics. Applied mathematics certainly goes outside of the Mathematics Department. But maybe artificial intelligence has remained even further outside. There is such a big range of hard problems. I'm not really qualified to talk about artificial intelligence.

I: There seems to be a gap between pure mathematicians and applied mathematicians. Some people think that it is due to the differences in philosophy and culture between them. How do we bridge this gap?

S: First of all, there has to be mutual respect. Even pure mathematics itself (or applied mathematics) is too big for anybody to bridge, much less to bridge all of mathematics. I think if we have mutual respect and cooperation and an open mind when looking for places to contribute, then the gap is not important. It is the solving of problems that is important. I feel that SIAM (which is an applied mathematics society) and the American Mathematical Society (which is more pure) have included everything, and they are cooperating more and more. I would expect that there is similar cooperation at the Institute here.

I: Should applied mathematicians have the same faith that physicists often have in their usually intuitive, *ad hoc* and non-rigorous methods in solving many of their mathematical problems?

S: You see physicists getting very clever, usually correct, results. There are definite differences between physicists and applied mathematicians. And probably many physicists are not as close to the computer as many applied mathematicians would be.

I: We shall round up with this question: Should mathematical rigor be imposed at the undergraduate level?

S: For teaching, I think maybe not. I am not sure if I want you to print that! The idea with examples is where students learn. My goal in teaching is always ideas and examples and not proofs. Of course, the mathematics has to be correct, but to prove that it is correct at every step, I don't see as important in teaching. Those students who naturally have that ability will go that way, but students who have other abilities should be allowed to go in the directions natural for them. That's my thought about teaching. If we want mathematics to be chosen by good students, to be liked by students and to be helpful to society, we need a broad view of our subject.

Eitan

Zen of Computational Attraction

Eitan Tadmor has made fundamental contributions to numerical analysis, the general theory of applied partial differential equations and scientific computation. His influence on applied mathematics is as deep as it is wide-ranging and as mathematical as it is organizational. His prolific research output, both personal and collaborative, must surely rank him as one of the top leading figures in his field. His direct influence may be glimpsed from an article written on the occasion of his 50th birthday and published in *Computational Methods in Applied Mathematics*, Vol. 4, No. 3 in 2004.

Tadmor's contributions span a spectrum varying from research to administration and institutional organization. He has given plenary lectures at many major scientific meetings, including an invited address at the International Congress of Mathematicians in Beijing in 2002. He was a founding co-director of the National Science Foundation (NSF) Institute for Pure and Applied Mathematics (IPAM) at the University of California, Los Angeles and is currently the Director of the Center for Scientific Computation and Mathematical Modeling (CSCAMM) of the University of Maryland, College Park. In addition to his being a professor concurrently in the Department of Mathematics, CSCAMM and the Institute for Physical Science and Technology at Maryland, he also holds the university's title of Distinguished University Professor. The list of his professional

Tadmor

services, whether on worldwide scientific committees or editorial boards of numerous leading journals in applied mathematics is vast, and clearly shows breadth and personal commitment rarely found in a single scholar.

When Tadmor visited the Institute as an invited speaker of the program on nanoscale material interfaces, he was interviewed on 11 January 2005. The following is an edited and enhanced transcript of the interview, in which he reveals his unusually early, if not precocious, fascination with mathematical analysis, both pure and numerical, and a Zen-like attraction to things computational and algorithmic. It also gives an insight into his consuming passion for research and total commitment to the scientific community.

Imprints: Could you tell us something about your graduate training and how you became attracted to your present research interests?

Eitan Tadmor: I was attracted to mathematics during the early years of my elementary school in Tel-Aviv. At age 13, I joined a local math club which was run by a Tel-Aviv University professor, Gideon Zwas. He had a lively personality and a great ability to make very appealing presentations of mathematical ideas. I attended Zwas' π-club throughout my early years of high school. It was there that I first became interested in applied and numerical analysis. Zwas became my first mentor as I had begun taking courses at the university while still in high school. Back then, in 1970, it was the first administrative arrangement of its kind. Later on, I was pleased to see how it paved the way for more established channels of bright students who wish to study an academic curriculum during their high school years.

In 1973, I continued with my graduate studies in applied mathematics at Tel-Aviv University. Tel-Aviv had an outstanding group of numerical analysts. Those were formative years for me, with teachers and students who later became colleagues and who helped shape my interest in applied and numerical analysis. I felt very comfortable with analysis and was caught up by the interplay between the analytical and computational aspects of numerical algorithms.

Later, I continued with my postdoctorate studies at CalTech which was home to one of the top-rated groups in applied and computational mathematics. It was then that I met Professor Heinz-Otto Kreiss, who later became my mentor. I was greatly influenced by his work. Fifteen years later, we ended up as colleagues at UCLA and have remained close friends over the years.

I: Is the π-club still in existence?

T: No, unfortunately. It lived through the '70s and it was very successful in attracting many bright young mathematicians, led first by Professor Zwas and later by Professor Moshe Goldberg. The topics covered in those weekly meetings were a cross section of analysis and computation. I know that the memory of the π-club remains very vivid in the minds of those who attended it; I know it is in mine.

I: You were attracted to numerical analysis right from the beginning?

T: Yes. My interest grew out of my education. There were the analytical tools that one learns about at the early stages. I mentioned my early years at the π-club and at Tel-Aviv University. At the same time, there were the numerical algorithms that one could implement and one would like to know if they work; more important, why they do not work. The analysis part of the numerical analysis plays an important role in clarifying these "if" and "why" parts. Often, these questions cannot be addressed within the numerical universe *per se*: they cannot be divorced from the underlying mathematical model they are trying to simulate. I have always liked the interface between mathematics and numerics. I still do. Back then, mathematics was used to design more efficient numerical algorithms. Today, there is a feedback loop, when numerical algorithms impact the kind of mathematical questions being asked.

I: Do you actually use the computer to create the patterns and algorithms?

T: Well, the computer cannot replace the creative process. So, the answer is "no". If you would like to create or analyze a numerical algorithm you are on your own. But this is only part of the answer. The computer is the test bed which enables your ideas to be tested. It is the new experimental laboratory, so you are not alone but you go back and forth. At the same time, the answer is "yes"; the computer, or more precisely its output, does produce patterns. These could be just spurious numerical artifacts; but they could also reveal new phenomenon that was not observed before and drives a new mathematical inquiry. It is a partnership. Numerical algorithms are an asset of this partnership.

I: Do you actually devise the algorithms?

T: Yes. I devised, for example, algorithms for solving certain nonlinear partial differential equations. Other algorithms were constructed which enable me to "manipulate" various representation of discrete data. You often read that the computer "solves" a problem. The computer does not solve anything. It implements different algorithms to solve different problems. In my case, I am interested in developing and analyzing numerical algorithms which produce accurate solutions for differential equations, or process voice and image data.

I: Am I right to say that you are not interested in the algorithms *per se* but that you are interested in algorithms to solve partial differential equations and so on?

T: You are right. I am more interested in the mathematical aspects of such algorithms. Having said that — numerical algorithms are not just a thought experiment; computers enable us to actually implement the algorithms we devise. This is the partnership I was talking about and the various aspects of implementation, therefore, become an integral part of the overall computational task.

I: Computer scientists study algorithms too. Are you interested in computer science?

T: The various aspects of implementation I was talking about are traditionally a main focus of computer science. But computer science can be viewed as the totality of what computer scientists are interested in, and there are many who study the algorithmic aspects combined with the mathematical aspects. There is no clear borderline. It is more a question of different emphasis. My emphasis is more on the mathematical aspects and less with the algorithm *per se*.

I: Do you consider yourself to be more of an applied mathematician than a pure mathematician?

T: I am an applied mathematician. But considering my work on the theoretical aspects of partial differential equations and their counterpart in numerical algorithms, some applied mathematicians would classify me on the pure side. It is difficult to decide where pure mathematics ends and where applied mathematics begins. The branches of mathematics I am involved with are primarily analysis and scientific computation, which, in the great vague divide of classifying pure and applied math, is categorized on the applied side.

I: Do you use the algorithms to actually solve the differential equations?

T: I develop algorithms for accurate solution of differential equations and I test them on a host of model problems. These then become tools that are applied to solve a host of problems in various fields. In some cases, the problems and the methodology for solving them could be very specific. For other cases, we developed a family of "black box" solvers which are portable enough to solve differential equations from a great variety of different applications.

I: What about the Navier–Stokes equations?

T: The set of Navier–Stokes equations governs the dynamics of flows at the human scale. That is, everything from air flow around airplane wings to the water flow in your bath tub. One might be surprised, maybe even worried, that just one simple set of equations is sufficient (or supposed) to describe so many different phenomena. Well, the Navier–Stokes equations are essentially one set of equations, worked out by the giants of the past. But they are not as simple as they appear to be. It is not totally surprising, therefore, that our mathematical understanding of these equations is incomplete. Indeed, there is a $1 million Clay [Institute] Prize for successfully clarifying part of the puzzle surrounding the mathematical quantities governed by these equations: what properties do they have? But even without the full understanding of their mathematical properties, we are developing numerical algorithms for the approximate solution of these equations. Practitioners compute the numerical solutions without necessarily waiting for their full mathematical understanding. At the same time, it excites a lot of research, a lot of ingenuity, and a lot of numerical experiments which try to complement each other in gaining insight into the mathematical properties of the Navier–Stokes solutions. A large component of the weather system, for example, is also governed by the Navier–Stokes equations. Here, interactions occur across several scales which are still human scales, say, larger than atomic scales and smaller than cosmological scales. The enormous complexity cannot be contained between the purely analytical walls, but it requires modeling and numerical simulations.

I: Can we say that in some sense, numerical analysis depends on the ability to design more and more powerful computers?

T: This depends on what your meaning of the word "depends" is (excuse the cliché). The mathematical models we are trying to simulate are independent of computers and to a large extent, so are the numerical algorithms which perform these simulations. The numerical analysis of such algorithms is intimately connected with the properties of the underlying model. In this sense, numerical analysis is independent of the computer running these algorithms.

At the same time, more powerful computers alter the kind of questions we may ask about our numerical algorithms, and lead to different notions of what optimal algorithms are. If in the past, it took 48–72 hours to simulate tomorrow's weather, then naturally, numerical analysis turned its focus on developing much faster algorithms. Over the years, the speed-up in computer power accelerated by Moore's law, doubling itself every 18 months. Everyone is familiar with it from his or her PC, but this doubling factor also applies to the new and improved numerical algorithms that were developed over the years. So nowadays, when it is feasible to compute tomorrow's weather in less than 24 hours, the demand arises to include more realistic models, or to develop new algorithms to include much better visualization. In this sense, numerical analysis depends on more powerful computers. Moreover, if computers become powerful enough, they can run different algorithms that communicate across different scales, and thus, instead of using mathematics to model the ensemble of small scales we can think of numerical algorithms performing the modeling "on the fly". Clearly, this requires the development of new numerical algorithms which are by-products of more powerful computers. A more powerful computer will enable us to reach new territories that have not been reached before. Still, you need the creative process of developing and analyzing new numerics to conquer these territories, and this is independent of how powerful the computers are.

I: Have the powerful computers raised any new issues and brought about new developments in numerical analysis?

T: Absolutely. The canonical example is parallel computers. Parallel computers completely changed the scope of what algorithms can do. What might be impractical or even impossible to do with computers based on a single processor becomes possible to do in parallel processing. Quantum computing could be the next and perhaps ultimate leap.

I: So the limit exists?

T: I think it does. One might say that if we just have powerful enough computers, many times more powerful than what we are having today, then we will be able to solve *everything*, to simulate a complete dynamical ensemble, perhaps even from quantum scales all the way up to the human scales. I do not know about the technological barriers here. But I will argue that even if the technological difficulties will be resolved, say in the next 100 years, still the smaller it gets, the slower the clock gets. That is, one needs to sacrifice a certain level of detailed information for having a computational algorithm to run at a finite time, that is, finite in human scales. And as powerful as computers can get by miniaturization in space, they will slow down in time. In this sense, limits exist and there is room for developing numerical algorithms for mathematical models which will bridge this gap of space and time.

I: Could you give us some examples of successful numerical algorithms?

T: The Gauss elimination method for solving N linear equations with N unknowns is perhaps the most ubiquitous numerical algorithm of all. It was always out there. But the Fast Fourier Transform was not and its discovery in the mid 1960s has had a lasting impact. It computes the periodic building blocks of general waves based on N samples and it reduces the computational work by order N. This is Huge. If N has the reasonable size, say, of ten thousands, then this is equivalent, by Moore's law, to 10 years of hardware speedup. These are exact algorithms. Their success is based on clever rearrangement of the computed quantities to achieve the final result in a fraction of the time it requires for a straightforward computation. In other cases, only approximate solutions are sought. This is the case with the solution for a host of partial differential equations drawn from various branches of science; the equations are just too complicated to be solved exactly. Approximate solutions are satisfactory. Here, there is a trade-off between how accurate the computed solution is versus how fast it can be computed. Many modern numerical algorithms are successful in making this trade-off. These numerical algorithms are successful in being very efficient.

I: There is an old perception that a mathematician's job is done once the model is formulated and that the rest is the job of the mathematical technician. How much has the role of the mathematician in modeling changed over the years?

T: On the contrary, the job of a mathematician just starts when the model is formulated. The modeling I am referring to is not necessarily mathematical modeling. Before the genome, there was the double helix and it was more descriptive than quantitative. Before Kepler's laws, you had Copernicus and his concept was not as quantitative but has had much more impact than Kepler's. Mathematics seems to be the most successful language to translate our qualitative concepts about the physical world around us into a set of quantitative statements. Today, more than ever before, there is a large effort to duplicate this success with quantitative biology. Still, the mathematical modeling is not left to mathematicians but to the interaction between scientists from different disciplines with mathematics. Once a model having its roots in biology, nano-science, chemistry or astronomy has been quantified, mathematicians study the interconnections, trying to fit the mathematical model as part of a greater puzzle. Often, modeling is a much more laborious and less glamorous task than the formulation of "$E=mc^2$". It involves experiments, measurements, statistical evidence and numerical experiments. More than before, pure and applied mathematicians are involved in all those aspects. This is particularly true with regard to the computational aspects. In the past, there was one critical reality check for a scientific theory, namely, that its predictions can be proved or disproved by experiment. Nowadays, computations provide another reality check for developing theories. This is the interplay I was talking about before, of numerical algorithms simulating mathematical models from different scientific disciplines. This is the intersection called "scientific computation".

I: Is there a coherent theory of scientific computation or is scientific computation nothing more than a collection of *ad hoc* methods and clever techniques?

T: Yes, there is a coherent theory of scientific computation. There are the fundamental concepts, technical tools, hierarchy of knowledge. But like every other area in mathematics, there are many isolated islands. Scientific computation, more than most of the other areas in mathematics (statistics might be the exception), is at the forefront of interaction with the other sciences and therefore, it has a constant flood of new input from the "outside". It absorbs clever new tricks, ingenious mathematical techniques and algorithms we do not always understand why they work so well. But over time, some coherence emerges. Sometimes, there are *ad hoc* methods pulling the theory forward. Other times, the theory identifies the danger zones where the numerics will not work, or even worse, will work out the wrong solution. There is a healthy tension between the hierarchy of knowledge and the collection of *ad hoc* methods. Over time, they merge into a coherent theory of scientific computation. It is a relatively young area of 20th century mathematics and it is a very lively one.

I: Do you consider mathematical modeling a science or an art? Is there some quality beyond mathematical expertise that one need to possess in order to be successful in mathematical modeling?

T: Mathematical modeling is an art, expressing itself in a quantitative language. Well, there is no recipe of how to make a mathematical model. It requires creativity, curiosity, ingenuity, imagination in addition to understanding the science behind the model.

Let me mention the example of modeling images. Images are all around us and nowadays our world is going digital. Digital images are collections of many pixels. My digital camera has five mega pixels. And each pixel has its own, very local grayscale (or color scale). We do not see these individual pixels, but instead, we see their collection as an image. In the last decades, many models were developed in order to manipulate the collection of pixels as images so that we can transmit, compress and in general, manipulate digital images. There are many mathematical models but there is still no final word about the one way that we should interpret a collection of pixels as an image.

I: Is the model independent of what goes on in the brain? Are the images not affected by the processes in the brain?

T: This is the reason I mentioned this example of image processing. Modeling digital images lives outside the mathematical universe. There is the mechanical part of the human eye. There is the conceptual part of the brain which puts together an ensemble of small pixels and gives them sense of what we understand to be an image. Once again, mathematical modeling seeks to match the world we see around us on the human scale, to a world made up from basic elements on a much smaller scale.

I: What is the greatest satisfaction you have had in your research career?

T: There are the moments you understand the answer to a mathematical question that bothered you for a long time. You know when you unlock the puzzle. These moments are very rewarding. It is a peak of a creative process. Another rewarding aspect is the development of numerical algorithms. It is rewarding to see a numerical algorithm that you have thought about, realized on the computer. And there is a great satisfaction in learning new ideas that are born into mathematics. There is a constant process of renewal, a generation of new ideas. There is still part of me which remains as excited about mathematics as I was during my days in the π-club. This is very satisfying. I feel blessed.

I: I believe that our Institute (IMS) is modeled in part on your Center for Scientific Computation and Mathematical Modeling. While IMS spreads its programs over diverse fields, your Center is extremely focused. Could you tell us something about your Center?

T: The Director of IMS, Louis Chen visited us in 2001 when I was the Director of the NSF Institute of Pure and Applied Mathematics (IPAM) at UCLA. This visit took place just before the IMS was launched. Certain aspects of the IMS, like its scope, covering a wide spectrum of mathematical areas, are modeled after the national NSF institutes such as IPAM, MSRI and IMA. In this sense, the IMS serves the purpose of appealing to a wider spectrum of the pure to the applied crowds. In 2002, I was recruited by the University of Maryland to serve as a Director of its Center for Scientific Computation and Mathematical Modeling (CSCAMM). Our Center is not a national center. It is a major initiative which is completely funded by university, devoted primarily to scientific computation and mathematical modeling and their interaction with other scientific disciplines. I elaborated before on my views as to the role of mathematics in this critical junction. A main part of our visitors' program centers around CSCAMM workshops. We organize several workshops each year and they have already achieved a considerable success in increasing the visibility of the outstanding faculty and activities in Maryland. Our center is an independent arm of the College of Physical Sciences in Maryland, with the mission of increasing the interaction between the different units through their common interface of scientific computation. Thus, the flavor of CSCAMM is somewhat different from the national institutes and it has a different strategic direction, as it is trying to lower the barriers between faculty inside and outside the university through the more focused platform of scientific computation and mathematical modeling.

Michael

Optimization, an Interior Point of View

Michael Todd is well-known for his fundamental contributions to continuous optimization, both in the theoretical domain and in the development of widely-used software for semidefinite programming.

His research work has left a deep impact on the analysis and development of algorithms in linear, semidefinite and convex programming; in particular, on interior-point methods, homotopy methods, probabilistic analysis of pivoting methods and extensions of complementary pivoting ideas to oriented matroids.

He did his BA at Cambridge University and PhD at Yale University. Except for a two-year stint at the University of Ottawa, his scientific career began and developed into prominence within Cornell University, where he is now the Leon C. Welch Professor in the School of Operations Research and Industrial Engineering.

Todd

He has been invited to give talks at major scientific meetings and universities throughout the world. He held special appointments at leading universities and centers of research in economics and operations research, such as the Fields Institute (Toronto), Carnegie-Mellon University, the Cowles Foundation for Research in Economics (Yale), the OR Center (MIT), the University of Washington, BellCore (US), Cambridge University and the Center for Operations Research and Econometrics (CORE, Leuven, Belgium). He has served, and continues to do so, on the editorial boards of leading journals on optimization, operations research and computational mathematics. Among the honors and awards given in recognition of his important research are Guggenheim and Sloan Fellowships, the George B. Dantzig Prize of the Mathematical Programming Society and SIAM, the John von Neumann Theory Prize of the Institute for Operations Research and the Management Sciences (INFORMS) and INFORMS Fellow.

Todd has close research links with NUS faculty in the Department of Mathematics and was Chair of the Organizing Committee of the Institute's program on *Semidefinite Programming and its Applications* held during 21 December 2005–31 January 2006. During his visit for this program, he was interviewed on 9 January 2006. The following is an edited version of the interview in which he gives us a stimulating glimpse of the theoretical insights behind one of the most important applications of the mathematical sciences to operations research, engineering, economics and industry.

Imprints: You did a BA in mathematics at Cambridge and went to Yale to do a PhD in administrative sciences. Was the thesis topic a mathematical one?

Michael Todd: Yes. I took a course from Herbert Scarf in mathematical economics at Yale. He described his recent work in computing approximate fixed points, and I got very fascinated by his work and, in general, by complementary pivot algorithms which use purely combinatorial arguments to solve optimization problems. I wanted to understand the combinatorial background to these methods. That was the basis of my thesis. It was indeed a mathematical one. "Administrative sciences" is a strange name. There aren't too many departments of administrative science, and they chose it so that it didn't sound too much industrial, too much business school. Basically, it's about the science and mathematics of decision-making.

I: Why didn't you go to the Mathematics Department instead?

T: I had been supported in Cambridge by Shell. They had a fellowship for me and they suggested that I go abroad for a couple of years to a business school. With a fellowship between my college in Cambridge and Yale, I went there mainly to see America for a couple of years and then I decided to stay because it was fascinating. Choosing the department was sort of difficult, and it was really an accident.

I: Were you interested in pure or applied mathematics right at the beginning?

T: At Cambridge, my work was basically in pure mathematics, but towards the end of it — and especially when I was at Yale — I decided that the applications were interesting. I got fascinated by the applications, in particular, by algorithmic questions.

I: Is semidefinite programming a generalization of linear and convex programming?

T: Semidefinite programming is a generalization of linear programming. In linear programming the variable is a vector whose components all have to be non-negative. In semidefinite programming, you have a symmetric matrix and all its eigenvalues have to be non-negative, so it has to be positive semidefinite. So it is more general than linear programming but it is a subclass of problems in convex programming.

I: Could you give us some examples of problems that involve semidefinite programming?

T: One of the nicest things about semidefinite programming is the wide range of areas in which it has been applied. I think that the first interest probably came from people in control theory who wanted to study ways of controlling dynamical systems optimally and making sure that they were stable. That led to inequalities that required certain matrices to be positive semidefinite. There are also applications in a completely different area related to combinatorial optimization problems connected with graph partitioning. Another source of semidefinite programming is robust optimization, which has been a hot topic recently. All of these different areas lead to an interest in efficient algorithms for semidefinite programming.

I: Is there an optimally efficient algorithm for solving linear programming problems?

T: That's the holy grail of linear programming research. It's a very intriguing situation. Now we have two different classes of algorithms — simplex algorithms and interior-point methods, and there is wide disparity between them on some classes of problems — sometimes one is much faster than the other. They are very different theoretically. The simplex method in the worst case is exponential but seems to perform very well in practice. Interior-point methods have a polynomial time bound and they perform much better than that bound in practice. For large-scale problems, it is not clear which one is the more efficient. There may be some new methods that will do even better. We're still waiting to hear about that.

I: Are these two methods connected?

T: Not very closely. They are based on very different geometric views of linear programming. The set of feasible solutions in linear programming is a polyhedron and the optimal solution always lies at a vertex. So it's natural to consider algorithms that just go from vertex to vertex and that's what the simplex method is based on — an algorithm that traces the skeleton of this polyhedron. Interior-point methods move through the interior and make smooth approximations. So they ignore much of the combinatorial structure and look at the analytic structure.

I: So one is discrete and the other is continuous.

T: Exactly. Interior-point methods never get an exact solution unless you do a special rounding procedure, but they get very, very close, and incredibly fast: you have to solve a very small number of systems of equations which are more complex than the equations in the simplex method.

I: How do you know which method to use?

T: It could be based on the software that you have. Most efficient commercial software allows you the option to use either one. I think people look at their class of problems and decide which one works better for their problems.

I: Which one is more popular?

T: I think for historical reasons the simplex method is more popular, but if you want something jazzy, the interior-point method is certainly a wonderfully efficient method for solving these problems.

I: Are there any probabilistic methods?

T: There are but we should distinguish two viewpoints. First of all, some algorithms make random choices and there are some very interesting theoretical ideas that have been used in low-dimensional problems that have much better computational complexity on certain classes of problems than the more usual ones. But there are also probabilistic analyses of the deterministic algorithms that people typically use on large-scale problems. Simplex and interior-point methods work in practice much, much better than their worst-case bounds. We would really like to understand that. One way to do it is to assume that the problem is random and to understand the average behavior of the algorithm on random problems. Some very interesting results have been obtained along those lines.

I: I noticed that there is a mention of homotopy in one of your papers. Is there something topological about it?

T: I think it is more a question of how the methods are based on different geometric views, and earlier I described a little bit how the simplex method is based on the combinatorial geometry and the interior-point method on the convex geometry. My earlier work was related to algorithms for computing approximate fixed points: homotopy ideas come up, but also the combinatorial topology and geometry of triangulations. Those algorithms were very interesting but not too much can be said about their computational complexity. They tend to be useful for small dimensions, up to maybe 50, on very nasty nonlinear problems, whereas linear programming and semidefinite programming problems are often much, much larger and more highly structured.

I: Is the software for implementing the algorithms freely available?

T: That really depends on whether you are talking about linear programming or semidefinite programming. Linear programming is very widely applicable and has huge commercial implications. So, the very best codes cost you some money, but there are some very good codes that you can obtain freely. There are a couple of websites where you can get some good codes for linear programming. But for semidefinite programming, the market is probably more in the scientific and engineering community; so you can't charge them a lot of money. Most of the algorithms are freely available, and several of those are available on the web. A good starting point is the NEOS Solver for Optimization site.

I: Have you written some of those yourself?

T: Yes, actually with one of my National University of Singapore colleagues and another colleague: Kim-Chuan Toh, who's in the Mathematics Department here, and Reha Tütüncü of Carnegie-Mellon University. We have a package for semidefinite programming, and it can also be used for linear programming.

I: Could you give us an idea of the complexity involved in semidefinite programming?

T: I'll give you some sort of an idea. First of all, these interior-point methods have been extended from linear programming to semidefinite programming. They typically take a very small number of iterations, perhaps 10 to 50, but each iteration involves a lot of work. Even if you have a problem with sparse data, in the semidefinite case you have to solve a generally dense large linear system of equations and that can be very costly. So these methods are typically very computationally burdensome, and the number of linear constraints can only get up to a thousand or two. These algorithms can give very accurate solutions. Other classes of algorithms, based more on first-order methods, can solve much larger problems with tens of thousands of constraints. They get much less accurate answers and don't have such good complexity bounds, but can be quite fast in practice. I'd say a thousand to ten thousand is the order of the matrices involved and the number of constraints that you can handle with these methods.

I: Can all linear programming problems be solved in principle by quantum computers or a theoretically most powerful computer?

T: I don't know a huge amount about quantum computers. From what I understand, I think it is possible to solve linear programming problems in one step. There's only a finite number of possible options, the vertices of the polyhedron, and the quantum computer is allowed to examine them all simultaneously and pick out the best. Similarly, for biological computers based on DNA and so forth. I don't know how practical these methods are. For semidefinite programming, I don't see that you can get an immediate solution, but it will be interesting to find out.

I: What happens if one day we really get quantum computers? Will linear programming problems be trivialized?

T: Yes, but maybe also all NP-complete problems too. It's not clear that these methods can really push all problems that are currently considered interesting to become totally trivial. I don't know whether such computers will really ever become that practical.

I: Do you believe in quantum computers?

T: I think it is a nice theoretical concept to consider, but I'm not expert enough on computers to comment on that.

I: Do you consider yourself to be an applied mathematician or a pure mathematician?

T: I'd say applied mathematician — that's what I say to people I meet on the plane. But just as with pure mathematics this generally gets the same response, "That was my worst subject. I don't understand it at all," which is very unfortunate. Sometimes I try to explain some of the nice things that mathematics can do.

I: Do you think algorithmically or geometrically?

T: I think geometrically a lot of the time. There are so many different ways of looking at optimization problems, from optimality conditions, to the theory of the algorithms and the modeling. I try to keep computational concerns in the back of my mind, but I'm still very interested in the theory as well. The geometric viewpoint on optimization problems really attracts me.

I: But at the end of the day, you still have to do the computations.

T: Yes, you do, and it's nice to be within, say, six degrees of separation, or fewer, from people who are actually practically solving applied problems. Even if you are not producing the software, you are motivated by improving the algorithms so that people can actually solve larger problems faster.

I: Except for two years in Canada, you have been at Cornell right from the beginning of your career. Have you ever thought of moving to other universities?

T: There have been a few times when I thought about it. But overall, Cornell has been a very attractive environment for me. The School of Operations Research and Industrial Engineering has some wonderful colleagues, both in optimization and more generally in operations research. The university as a whole, and mathematics and engineering, have wonderful people, and the quality of the graduate students has been terrific. I really enjoy working with the students in operations research and applied mathematics. It's also a wonderful place to live and very naturally beautiful.

I: Do you talk to people in economics?

T: Economics, once in a while, probably less than people in engineering, computer science, mathematics, but still occasionally, yes. My interest in economics was more during the '70s, a long, long time ago. I have sort of lost touch with the latest things that have been done now.

I: What advice would you give to a graduate student who is interested in applied mathematics?

T: You really need to find a problem where you feel so excited about it that you have a fire in your belly to keep working on it. You should look at all options, keep your options as open as possible, find an advisor to help you see the right approach at the right time and to let you do what inspired you to work in the area, and hope you find the way ahead of you.

I: Have you gone back to Britain?

T: I've gone back socially, for family reasons or whatever, every year or so. I spent one sabbatical back there and I've been back for several conferences. I find in the area I'm working in there are interesting people in many places in the world: in England, but also in Belgium, France, Germany, Japan and Singapore besides the US that I work with as well.

I: The name of the school you are in — "School of Operations Research and Industrial Engineering" — seems to give people the impression that it has very little to do with mathematics.

T: It's more of a question of how it evolved. We have people who are much more involved with practical work and consulting, but I think many of us regard ourselves as a mathematical sciences department within engineering. We have people working in applied probability, statistics, and optimization, from quite a theoretical viewpoint to a more practical viewpoint. It's nice to have that full spectrum, but many of the faculty were very well-trained mathematically. A lot of us have appointments also in the Center for Applied Mathematics, and some people have appointments in mathematics as well.

I: How is your relation with the engineers?

T: Pretty good. Some fields of engineering are closer than others. We are not too much involved in the experimental side, but for example, our relations with electrical engineering and computer science are very good.

I: Do you try to educate the engineers mathematically?

T: I try. I very often have students from other parts of engineering taking my classes. Along with the modeling and computation involved, I try to make them understand that the abstract viewpoint can be valuable. I hope they appreciate the beauty of mathematics. I think that in a strong engineering college, the students are pretty much aware of the advantages of having good mathematical training, particularly, the graduate students.

Sergio

Wireless Communications, at the Shannon Limit

In the beginner's mind there are many possibilities, but in the expert's mind there are few.
– Shunryu Suzuki (1904–1971), Zen Master

Sergio Verdú is world-renowned for pioneering the field of multiuser detection in wireless communications and for fundamental work on data transmission and compression in information theory.

His theoretical doctoral research has a tremendous impact on communications technology with numerous applications in mobile cellular systems, fixed wireless access, high-speed data transmission, satellite communication, digital television and multitrack magnetic recording. His book, *Multiuser Detection,* published in 1998 is now a modern classic. His research papers have received many awards from scientific and professional bodies. He has also received several awards for professional education and outstanding teaching. The prizes, awards and accolades bestowed on him are indeed too numerous to list; the latest in 2007: election to the US National Academy of Engineering and the Claude E. Shannon Award, the highest honor in information theory.

On the faculty of Princeton University since 1984, Verdú is Professor in the Department of Electrical Engineering since 1993. He is also a core faculty member of the Program in Applied

Verdú

and Computational Mathematics. He is known for his personal zeal in advisory and organizational work (not only in the United States, but from South America across Europe to Asia) in advancing and promoting science and technology. He has served as President of IEEE Information Society and he serves on the editorial boards of leading journals in his field, in particular, for IEEE. His scholarship and scientific charisma have led to many visiting appointments and invited lectures around the world.

Verdú was an invited speaker at the Institute's program *Random Matrix Theory and its Applications to Statistics and Wireless Communications* (26 February–31 March 2006). He was interviewed on 26 February 2006. The following is an edited and enhanced version of the interview in which he traced his scientific path from humble beginnings in Barcelona, Spain to prominence in the world's leading centers of communications research in the United States. It resembles a classic Spanish narrative that spans a wide spectrum from human passion to intellectual vision set on a scientific stage for exploring the physical possibilities of communications at the edges of the theoretical limits of information theory.

Imprints: In your undergraduate training in Spain, you had already specialized in telecommunication engineering. Why did you choose this particular branch of engineering?

Sergio Verdú: I decided to become a telecommunications engineer when I was seven years old. My father gave me a toy — a kit with which you could build radios and all sorts of electrical devices — and I was hooked. My father was very good with electrical gadgets. As a child, I was always immersed in electronics. At that point, I decided to be an electronics engineer and I never wavered from that.

I: Was your father an engineer?

V: No, my father had very little formal education. His childhood was spent during the Spanish Civil War. He suffered a lot and went through tragic circumstances. He was a self-made man, a very fine man. He really had a lot of influence on me even though he died in an automobile accident when I was 11 years old.

I: What attracted you to go to the United States (in particular, the University of Illinois at Urbana-Champaign) for your graduate studies?

V: Going back to my early youth, I guess that's where you would find the traces for all these decisions. My parents decided that I would get an English tutor when I was six or seven years old. From then on until when I was in high school, I had an English tutor: a Spaniard, she was not a native English speaker and I guess that accounts for my less than perfect accent. It gave me an edge over everybody else who was just learning English in school. I was fascinated with things American and, in particular, with the space program. When I was 14, I jumped at the opportunity to spend a summer in Rockville Center, Long Island in 1973. It was an excellent opportunity to see a world completely different from the backward country where I had grown up: Spain was under a fascist dictatorship from 1939 to 1975. At that time things like color television were completely new to me. I remember being fascinated by the Watergate affair that was going on at that time. The fact that a country could be so open politically while undergoing a painful episode and still able to do it with a sense of humor was a revelation. Although I tried, practically the only American activity that to this day I never could really get interested in was baseball. At that time, it was certainly not very common for Spanish students to do their graduate studies in USA. In fact, I didn't know anybody who had done that. I ended up at the

University of Illinois at Urbana-Champaign partly because as an undergraduate I had worked a lot on computer-aided design. I was very much into the design and analysis of electrical circuits using the computer. A lot of prominent people in that field have been at Urbana-Champaign. I was also admitted at Stanford and, of course, I knew some of the professors there but perhaps not as much as the ones at Urbana. But at that time, I already knew that I had done enough programming and hacking in my life. I really wanted to do theoretical work and I really wanted to do communications theory and information theory. Of course, as an undergraduate I had already heard of Shannon, and one day when I was discussing my options to go to graduate school in the US, one of my professors said, "Well, you know, Claude Shannon was at the University of Illinois." I said, "Oh, okay." That clinched the decision for me.

I arrived in the United States in 1980 during the Presidential campaign between Ronald Reagan and Jimmy Carter, and Urbana-Champaign was pretty shocking to me. It was so unlike the atmosphere in the New York area that I had seen in '73 and, needless to say, very different from the big European city life that I had been exposed to. In addition to the geographical isolation, the religious atmosphere of the place was really striking. When I got there, I asked, "So when did Claude Shannon teach here?" and nobody knew about Claude Shannon having been there. One day, browsing in the university bookstore, I picked up a copy of Shannon's *The Mathematical Theory of Communication*. It had been reprinted by the University of Illinois Press.

I: Then you went to Princeton immediately after Illinois?

V: Yes, the day after I defended my PhD thesis. My wife Mercedes and I drove our Chevy to New Jersey.

I: You didn't go back to Spain?

V: No. I always wanted to remain in the United States. I had a Fulbright Fellowship. That gave me a lot of trouble because Spain, concerned about the brain drain, refused to give me permission to stay in the United States. But, after a long, complicated process through the State Department and the Department of Justice involving senators and so on, I was granted a waiver of the requirement to return.

I: Your doctoral research pioneered the field of multiuser detection. Could you tell us something about it? Were you excited and surprised by your work at that time?

V: Yes. At that time (in the early '80s), I had worked for my master's thesis in minimax robustness. This was a field that originally started in statistics with the work by Huber in the '70s. Then there was a lot of work in engineering (particularly by my advisor Vincent Poor) applying Huber's theory to robust estimation, robust detection and so on. Vincent Poor mentioned that in spread-spectrum communications, they were modeling the multiaccess interference as white Gaussian noise, and although this seemed to be a pretty good modeling assumption, perhaps there was some room to apply robust statistical methods to account for the deviation from the central limit theorem. I started to look at it from that angle, but then I quickly realized that that was not the right approach and that a completely new approach had to be taken. Then I obtained the optimum multiuser detector, and that became the beginning of my PhD thesis. The interesting thing was not only the structure of the receiver but the fact that in many cases you could achieve single-user performance. The gain was remarkable and much more than what we expected. That was the beginning of multiuser detection. At that time, nobody was paying any attention to it. Spread-spectrum research was pretty much dominated by military funding and did not have the vibrancy it acquired later on, thanks to the ascent of wireless telecommunications and CDMA wireless commercialized by Qualcomm.

I: Your doctoral work was not classified?

V: The university would not allow any classified research. It was actually good for me that at the beginning it did not attract any interest. It was only years later that multiuser detection became a very vibrant research field with a lot of research citations to the early work I had done in the early '80s. In 1998, I published a book, essentially a compilation of my work and my teaching of the subject. But sometime in the late 1980s it ceased to be my primary research focus. Perhaps if the success had been immediate, then I would have devoted a lot of my efforts into that and less into information theory, which eventually became my primary field of interest. I think it was propitious, and interestingly, the time constant from inception of ideas to implementation of these ideas in that particular field was very, very long. It's only recently that there has been motivation and success in industry implementing multiuser detection. One of the drivers has been multi-antenna systems where there is interference between signals transmitted by different antennas. Qualcomm, the proponent of CDMA cellular wireless, came up with a second-generation cellular wireless with rather old signal processing algorithms. It didn't use any multiuser detection, but they have announced recently that they are using these methods in their third-generation chips. These are channels where bandwidth and power are resources to be conserved. And one of the lessons that Shannon taught us is that you have to exploit the fine details in your model (in this case multiuser interference) to squeeze the most out of the channel resources.

I: It is now commonplace?

V: It depends on which area. Although the systems were not designed with the idea that you would have sophisticated receivers taking into account multiuser interference, both in third generation CDMA and in Digital Subscriber Loops (high speed data through telephone copper wires), they are starting to implement it. Multiuser detection is commonplace in the multi-antenna receivers where you can get substantial gains in capacity taking into account interference proceeding from different antennas. There are also chips that take into account intertrack interference in magnetic recording. It is always just a matter of time until the maturity of technology puts a stop to the waste of bandwidth/power.

I: Have you ever considered working in industry?

V: No, I was always very much an academic type. I like the freedom to pursue my own ideas and my own work. I also like to interact with young people. Not having a boss is nice too.

I: You seem to be equally comfortable with mathematics and engineering. How do you manage to reconcile their two different approaches to problem-solving — approaches that are apparently poles apart?

V: Strangely enough, they are not very different because the way you approach problems is essentially the same in both fields: going back to the basics. As much as I can, I always try to avoid carrying a bag of tricks that I can apply from one problem to another. I have actually moved quite a bit from problem to problem, and there's a lot of pleasure starting on a problem from scratch that I really didn't know anything about. Like the Zen philosophy says, in the mind of the beginner the possibilities are endless. A lot of important contributions are made by people who have just entered the field. Learning new mathematics is a delightful reward. Technology points out what next to learn; for example, my work on random matrices — which is why I am here now — was motivated by wireless communication systems. The type of research that excites me is mathematically challenging and relevant to the real world. Claude Shannon was the archetypical seamless combination of mathematician and engineer.

I: Do you think that, in general, engineers have as much mathematical training as they should have?

V: Mathematical training is like wealth, nobody has enough of it. The thing about this discipline that we call electrical engineering is that its unifying theme (electricity) goes back to the 19th century and is now completely obsolete. But our engineering training gives you a lot of versatility to deal with very different problems. To give you an example, two of my graduate students are finishing their PhDs in information theory this summer and are joining Goldman Sachs and Credit Suisse. Electrical engineering undergraduates may not get as much mathematical training as they would need to be professors doing research on, say, telecommunications. That mathematical training they will have to get later on in graduate courses and on their own. But electrical engineering undergraduates do get very strong training in problem-solving, and this gives them a lot of options.

I: It seems that one perception about the mathematical training for engineers is that they are more interested in sort of recipes or a bag of tricks for solving problems.

V: The training in engineering is very different around the world. Some of the European systems tend to have a kind of dichotomy. In the first two years of engineering, they are very mathematically oriented, and then later the subjects become very practically oriented. For example, I had to take two semesters of television — something that would be completely unheard of in the US. I think that in the US, perhaps because the professors are much more research oriented than in other places, we tend to be more mathematically oriented, at least those of us on the applied mathematics side of electrical engineering, like communications, control and signal processing.

I: Do you work directly with hardware engineers to create the technology?

V: No, not really. By the way, the dichotomy between hardware and software is fading. It is always important to be aware at any given time what the technology can deliver so that you know whether the solutions you are coming up with are solutions that can be implemented now or in 20 years' time or perhaps the technology in a certain field has progressed so much that you can implement things that are much more sophisticated than what people are implementing right now. So it's very important to have a sense of what technology can deliver even if normally we don't collaborate in research with people working in hardware.

I: What is the "biggest" unsolved theoretical problem in communications technology?

V: The biggest success story of Shannon's theory has been in point-to-point communications. Shannon's theory has been instrumental in anything that has to do with modems, wireless communications, multi-antenna and so on. But network information theory has proved to be a particularly tough challenge. Shannon was the first to formulate the problem, or at least the building blocks, in 1961. Instead of having one transmitter and one receiver, you have a bunch of transmitters and a bunch of receivers, and you may also have some nodes in between that act as relays, and some of those nodes may also be sources or sinks of information. You could think of a very general topology and you would like to know what are the best rates of information, what are the distinguishable signals that you can send. This is something that we still don't know.

Another important technological challenge is data compression of audio and video signals, which in my view is still in its prehistory. Even though Shannon also gave the fundamental principles of this discipline, information theory has not had nearly as much impact as it has had in channel transmission or in text/data compression. I think the reason is that we do not yet have a good understanding of human vision and hearing, and even the little we know is hard to marry with the available theory.

I: The point-to-point problem is solved?

V: We understand it a lot better. Shannon gave us the point-to-point framework, but he didn't give us all the solutions. Finding the capacity of a particular point-to-point communications channel may be extremely challenging and, in fact, the capacity of some very simple channels is still unknown.

I: Do you agree that engineers are very focused in their research in the sense that they try to solve only problems that are of immediate practical concern in contrast to physicists who try to answer fundamental questions that are not immediately applicable?

V: No, I don't agree. Shannon was the primal example of an engineer who would explode this myth. Many of us who are working in theory are accused, more often than not, of doing exactly the opposite: of solving problems that are of no immediate practical concern and that may become relevant only in the distant future or never. Those of us who have followed in Shannon's footsteps have an appreciation for beauty and elegance and for the fact that beautiful and elegant results sooner or later become practical. So you need to have some faith even though what you are working on now is not of immediate practical concern. You may be interested in it not because of some technology out there clamoring for solution, but because of its beauty.

I: It appears that you are a mathematician first and then an engineer.

V: I would say first an engineer, then a mathematician and then an engineer. I have come full circle. My doctoral thesis had an important component in developing algorithms, and also a lot of analysis but I had this nagging feeling that it was not mathematical enough for my taste. When I got into information theory I became quite theorem-proving minded. But, the thrill of coming up with new algorithms is something I have come to appreciate later, more so in recent years. When I was younger, I had the idea that if I cannot prove a theorem about something, then I don't want to do research on it. Now my outlook has evolved. Of course, I still like to prove theorems, but I have also done some recent work that is algorithmic and I enjoyed it very much.

I: Can you tell us something about your present research interests and the problems you are working on?

V: People say that the interesting problems are at the boundary between disciplines. This is actually true sometimes. One of my current interests is the boundary between information theory and estimation theory. A couple of years ago, we found a very basic formula that connects some basic quantities from information theory and estimation theory. Capitalizing on this formula, we gave some simple proofs of a probability theory result on the monotonicity of the non-Gaussianness of the sum of independent random variables as well as a famous result from Shannon's 1948 paper, called the *Entropy-Power Inequality*. We also came up with a new universal formula in continuous-time nonlinear filtering, as well as an algorithm to minimize transmitted power. All those come from this innocent-looking formula. As usual, there is nothing more insightful and practical than a pretty formula.

Random matrix theory has been very rewarding. I got into random matrix theory around 1997. When I was finishing my book, I was fortunate to become acquainted with Marchenko–Pastur's theorem and I included it in Chapter 2. Since then there has been an enormous interest and excitement. It is challenging to get into this theory. Even though its early history developed in the '50s and '60s, the core results are quite recent. It's only in the last 10 years or so that there has been a lot of interest in it from contemporary mathematicians.

I: Is random matrix theory applicable in engineering?

V: Very much so. It's applicable and fundamental in wireless communications. The first application was in the capacity of multiple antenna systems. In Bell Labs, Foschini and Telatar realized that in the presence of electromagnetic scattering when you have multiple antennas the channel capacity can be much larger than if you have single-antenna transmitter and receiver. Random matrix theory is fundamental in this realization, and also in the analysis of the fundamental limits of spread-spectrum in wireless communications.

I: Have there been any breakthroughs in random matrix theory?

V: The great breakthrough, at least for applications in wireless communications, was in 1967 in the work of Marchenko and Pastur in the Soviet Union. That was an amazing piece of work. It was completely unknown for many years. In 1986, I looked at a random matrix problem that I wanted to solve. I looked in the literature (of course, that was before Google) and I could find nothing. The work on random matrices that I could find was completely orthogonal to what we needed. Physicists and mathematicians were rediscovering the Marchenko–Pastur result in the late '80s and '90s. Lately, there has been a lot of excitement in a new mathematical field called "free probability" and one of its main applications is in random matrix theory. Wireless communications and information theory have been one of the main propellers of work in this theory. We are not just consumers of this kind of result; we have also been able to pose new questions and solve some of these problems.

I: So, in a sense, wireless communications has affected the development of random matrix theory.

V: Oh, yes, for sure. You see this pendulum of interaction in other fields. Information theory was very much influenced by ergodic theory, and also the other way around. Kolmogorov made a fundamental discovery in ergodic theory thanks to information theory.

I: Do you do much consultation work for industry?

V: I occasionally have done work with people in research labs such as Bell Labs, Hewlett Packard, and Flarion, which was recently acquired by Qualcomm. When I was doing work in Hewlett Packard, the group there was very theoretically inclined. I was actually kind of like the guy who was pushing for us to do more algorithmic work rather than theorem-proving. It was particularly rewarding to be associated with Flarion because you see the thrill of seeing brilliant ideas being implemented in a very short period of time.

I: Do you have any patents?

V: Traditionally, being more academically oriented towards peer publication, I have not pursued patents at Princeton. But yes, I do have quite a few patents granted or pending both through Bell Labs and Hewlett Packard.

I: Do you think that technology will be able to catch up with the theoretical advances in science and technology or even mathematics?

V: Well, Shannon's theory is a good example of a theory that at the beginning created a lot of enthusiasm. Shannon became an instant celebrity. Then, for a few years, people were asking the question, "If this is so good, how come it hasn't seen the light?" Of course, what happens is that it came well before its time, well before technology was ripe to be implemented. It took a long, long time for implementable codes to achieve Shannon's limits. In data compression they did not appear till the '70s and in data transmission until the '90s.

I: When was Shannon's theory put forward?

V: 1948, so it took a long time. That's a powerful lesson because everybody knew that these were very powerful ideas. For decades, there was a lot of unsuccessful work in trying to design codes that would approach Shannon's limits. When there are theoretical breakthroughs and when we are able to solve problems of a fundamental nature, then just because technology doesn't seem to be on the near horizon to be able to implement those ideas or formulas, it doesn't mean we should give up and say, "Okay, this is a dead field because we have given it long enough time and technology has not implemented it, and therefore it is hopeless." I think information theory is a great lesson in that respect. In communications, we have a limited piece of spectrum that we can only use with given resources, and there is an enormous economic incentive to use that spectrum as efficiently as possible. So when you have a theory like information theory that sets fundamental limits, there is an enormous incentive to get as close as you can to those limits.

I: How would one make the theoretical work drive the technology faster? Is that possible?

V: Well, it is possible. In certain areas, it has been enormously successful, for example, in modems, in work that was published in the *Information Theory Transactions*. Four years later, you could buy modems that were implementing those ideas for a hundred dollars. That is a field where the time constant is much faster. In fields like cellular wireless, the technology transfer has been a lot slower. Developed in the late '80s, second-generation wireless systems were predicated on technology that was really old (a lot of it '50s, '60s). A revolution happened in the 1990s with the advent of a class of channel codes called the turbo codes. In the beginning, they were not very appealing to the theoreticians because these codes came very close to the Shannon limit, but nobody could explain why. We couldn't come up with theorems that would say, "Hey, of course, this is why they do work." Now, we understand them better. Actually, they vindicate Shannon because he came up with a theory for what the best code could do without the benefit of knowing a single code except possibly for the simple Hamming code that was developed at the same time. He said, "Well, I don't know how to construct the optimum code but I can show that a construction where the codes are chosen blindly at random, performs close to optimum on the average." The problem is that if you choose a code at random, it cannot be implemented because it doesn't have structure. So these new codes that go back to the 1990s turn out to have enough structure that you can implement and encode them in linear time and at the same time they have enough randomness like Shannon originally said in 1948 to be close to the best.

I: So without those codes the cellular revolution would have been impossible?

V: The first digital systems used codes that were really far from capacity. Early in the game in the design of codes, people took a turn away from Shannon's random codes. Coding theory became a geometric discipline, very combinatorial, not so probabilistic. Now we are going back to the roots. Those geometric constructions that emphasize minimum distance properties of codes are not the ones that achieve capacity. They are very interesting mathematically but are not the ones that turn out to come closest to Shannon's fundamental limits. With the new codes, you can increase the efficiency quite a bit. What is surprising is that there was nothing inherently there to prevent people in the 1960s to come up with these codes. Actually, Gallager at MIT had come up with random-like codes in the early '60s but he abandoned them because they thought they could never be implemented and that they were too complicated. They are actually not too difficult to implement. The key is not to attempt optimum decoding because that is too expensive. With a judicious choice of code, the life of the decoder is a lot easier, and near optimal decoding is feasible. The bottom line is that in linear time you can come very close to the Shannon limit. People are now using cellular phones that incorporate these codes.

I: Is the Shannon limit a real physical Heisenberg-type limit or is it a Gödel-type logical limit?

V: The short answer is: information theory is a chapter of probability theory, which in turn is a chapter in mathematics. The starting point is a stochastic model for the information source and a stochastic model for the channel. Are those models relevant to the real world? If they weren't, your cellphone would not work. Having said that, since 1948, there have been enormous strides in information theory dealing with uncertainty in nonprobabilistic ways. An example is the theory of algorithmic complexity which is devoid of any probability, and was put forward by Kolmogorov, the father of modern probability theory.

I: Do you think that a revolution in wireless communications would follow in the wake of breakthroughs in nanotechnology?

V: The radiofrequency spectrum usable in wireless communications is rather limited. Information theory tells us the fundamental capacity of the medium. We cannot go beyond it no matter how fast the computing technology. But let me address the question from a broader perspective: why can a DVD contain a lot more music than a CD? The compression technology of the CD dates back to the 1930s. By the time the DVD was developed 15 years after the CD, lossy compression was much better understood, and the optical recording devices were also quite a bit more advanced. So the engineer reaps benefits from both applied physics and applied mathematics. For the information theorist, new physical devices mean new communication channel models, with a capacity to be discovered. So I think information theorists are going to be around for a long time.

Michael S.

Breathing Mathematics into Genes

Michael S. Waterman is world-acclaimed for pioneering and fundamental work in probability and algorithms that have tremendous impact on molecular biology, genomics and bioinformatics. He was a founding member of the Santa Cruz group that launched the Human Genome Project in 1990, and his work was instrumental in bringing the public and private efforts of mapping the human genome to their completion in 2003, two years ahead of schedule.

After his PhD in statistics and probability from Michigan State University, he taught at Idaho State University and visited Los Alamos National Laboratory for a short period before going to University of Southern California (USC) in 1982 to pursue a long and distinguished career in molecular biology, mathematics and computer science. The well-known "Smith–Waterman algorithm", which he developed with Temple F. Smith in 1981 for determining the degree of similarity (homology) of amino acid sequences from DNA, RNA and proteins, is catalytic in igniting the bioinformatics revolution. The formulae, which he and Eric Lander derived in 1988, are crucial for the so-called

Waterman

shotgun strategy for assembling genome sequences by cutting up the genome into short fragments that are easier and faster to sequence and then fitting them correctly together. In 1995, he published the first textbook, *Introduction to Computational Biology: Maps, Sequences and Genomes*, that laid the foundations of the new field of computational biology, of which he is considered to be the founding father. When he first went to USC, he started one of the world's first cross-disciplinary programs connecting genetics, mathematics and the information and computer sciences. With the setting up of the Center for Computational and Experimental Genomics in 2001, Waterman and his collaborators and students continue to provide a road map for the solution of post-genomic computational problems.

For his scientific contributions, he was elected fellow or member of prestigious learned bodies like the American Academy of Arts and Sciences, National Academy of Sciences, American Association for the Advancement of Science, Institute of Mathematical Statistics, Celera Genomics and French Acadèmie des Sciences. He was awarded a Gairdner Foundation International Award and the Senior Scientist Accomplishment Award of the International Society of Computational Biology. He currently holds an Endowed Chair at USC and has held numerous visiting positions in major universities.

In addition to research, he is actively involved in the academic and social activities of students as faculty master of USC's International Residential College at Parkside.

Waterman has served as advisor to NUS on genomic research and was a member of the organizational committee of the Institute's thematic program *Post-Genome Knowledge Discovery* (January–June 2002). During one of his advisory visits to NUS, he was interviewed on 7 February 2007. The following is an edited and enhanced version of the interview in which he describes the excitement of participating in one of the greatest modern scientific adventures and of unlocking the mystery behind the building blocks of life.

Imprints: Your PhD was in probability and statistics. How did you get into biology?

Michael S. Waterman: My PhD thesis was in probability and I did my initial work in probabilistic modeling and iteration of deterministic functions. I got into biology in connection with Los Alamos. Stan Ulam, who was a mathematician, was interested in what mathematics you might need in the new biology. He brought Temple Smith to Los Alamos for a number of visits. Another scientist at Los Alamos named Bill Beyer had an NSF project for one summer and I came to work with Bill and Temple. That was how I met Temple Smith and what really started me in this area.

I: Ulam was not really a biologist.

W: Not at all. He started as a completely pure mathematician in the Polish school of mathematics famous for its problems begun in a café [Scottish Café]. He came to the US — I forgot who really brought him to Los Alamos — and worked on the Manhattan Project. He actually flowered there and contributed to all of the amazing crowd there. He was one of those few great men who was still around many years later.

I: It seems that he had some foresight and could see that biology would be a thing of the future.

W: Yes, but I don't think he knew exactly what it was. He saw that it was intriguing and different, and it was clear to him there was something there.

I: I remember he wrote a book *What is Life?*

W: That book was by Schrödinger. It's not really a very accurate book about the subject, but it inspired many people to take on the mathematical and other aspects of biology.

I: After Los Alamos, where did you go to?

W: When I started this work, I was a faculty member of Idaho State University. I was just visiting Los Alamos in the summer. Then, in 1975, I went there permanently until 1982. In 1982, for most of the year, I had a visiting appointment in the Department of Biochemistry and Biophysics at the University of California in San Francisco — a very biological place. Then I went to USC [University of Southern California].

I: All this while, you were still doing mathematical work?

W: I do mathematical work. At Los Alamos, I was in a statistics group analyzing energy data. Beginning with Temple Smith, I also worked on stratigraphics in mathematical geology. I worked on a number of different applied problems, but because of the connections with Bill Beyer and Temple Smith, I was doing some of this work in biology, mostly algorithmic, in biosequence metrics as a hobby until finally, about 1981, 1982, I decided that that was what I would really like to concentrate on.

I: Could you tell us how you got into collaboration with Eric Lander that led to those famous formulae in physical mapping?

W: Eric Lander was still in the Business School at Harvard when he became interested in biology. After that, he was at MIT and the Whitehead [Institute]. Eric had written a proposal to a private foundation and that foundation asked me to look at the proposal. I met him in that connection and so I knew him. He became more and more interested in molecular biology; initially it was going to be neurobiology. We had an acquaintance and our joint paper came at the beginning of the genomics revolution. At that time he had an office at the Whitehead Institute [for Biomedical Research]. I was visiting and a copy of PNAS [*Proceedings of National Academy of Science*] was on his desk, it had to do with the first papers on physical mapping, including a paper on physical mapping of yeast by Maynard Olsen, and John Sulston had a paper on physical mapping of C. elegans. The progress reported by these papers was slower than what people expected and the C. elegans paper, which used a different method, had a simulation study. Eric and I thought we should be able to do something with it. So we started thinking about the problem and I realized that it was a coverage problem. I remembered a little book I had seen — *Geometrical Probability* by Herbert Solomon — on coverage. We went to the MIT library in the middle of the night. We found the book but it didn't really pertain to the kind of problems we were looking at. Actually the problem we worked on was about car parking. Of course, you have a long street on which you park cars at random locations, allowing parked cars to overlap. What is the distribution of the coverage of the long street by this process? We found it was easier to work on the problem directly and we published our first paper. And then later on, Eric proposed a problem that was a little bit different and involved using short unique DNAs to anchor the clones (cars!) for the coverage. There was a paper by Arratia, Lander, Tavare and myself. Arratia and Tavare knew that the Poisson process was the right way to look at the problem. I really learned a lot from these guys.

I: I think Eric Lander once mentioned that he left mathematics for economics first and then biology because he thought that mathematics was too "monastic" and cloistered. In contrast, do you think that you are a mathematician at heart?

W: I went to USC holding tenure at the Math Department with a joint appointment in biology. Now it's the opposite; actually I hold a tenure in the Biology Department, but in my heart I'm still a mathematician and statistician. To pick up Eric's comment, I think there are certain people in mathematics who will be motivated to work in other fields. I'm certainly one of those. I like working on problems that require different strategies. One of the appealing things about working in biology is that there is more than one person interested in what you are doing. There is a community aspect of it which is very important to me and keeps me going. There is a very wonderful aspect to working in teams. I think at my age I would find it a very hard time continuing to try to prove things in ergodic theory had I stayed in that area.

I: In contrast, pure mathematicians tend to work by themselves.

W: By themselves, with the door closed. I can do that, but I kind of like talking ideas around and even with people telling I am wrong.

I: When you wrote the first textbook, published in 1995, that laid the foundations of computational biology, did you expect those dramatic developments in the field to occur within the following five years?

W: At that time, I didn't think we would have the human genome sequence by the year 2000. I knew there was going to be great progress, but things happened that I didn't expect.

I: When you wrote the book, has the project already started?

W: The first discussion, which I was lucky enough to be part of, was in Santa Cruz in 1985. There were a dozen people and I was the "computer person" in the group. In fact, there was a California magazine which had an article about this. In 1985, it seemed to be feasible, the computations seemed just possible. Someone brought up the cost of a military ship, and that put the cost of the project in perspective. The project started in 1990 and it was planned to take 15 years (or more) to finish it and push it further.

I: Was the book to some extent motivated by the project?

W: In part, by the data. I had been teaching a course since 1983. While teaching it, I was writing the book and correcting the chapters. I was trying to write the book for several years.

I: Was there any book before that?

W: There was a book [edited] by David Sankoff and J. B. Kruskal [*Time Warps, String Edits and Macromolecules*]. It was a book based on dynamic programming with various applications to genome sequences. That was an important predecessor. There was a book, a rather naïve book, looking at information theory approaches by Lila Gatlin, which was published in 1972.

I: Which is more crucial to the theoretical techniques used in gene mapping: the "better" algorithm or the "faster" computer?

W: I think, both, especially with the difficult mapping problems people attack today, not finding a mutation in a single gene but studying a complex phenotype involving multiple genes. It's not clear how far we will get. There are other extremely important computing facilities of which you have to take advantage.

I: Like parallel computing?

W: That's everywhere today, there is no doubt. In fact, students we recruit to our program ask about what computational facilities would be available to them.

I: Is it possible to break up the problem into small parts to work on?

W: Certainly for some problems it is possible to do it that way. Most of this parallel computing activity in biology is not sophisticated, but for problems where the processors have to communicate it is much more difficult.

I: What are the prospects of quantum computers in genome mapping in the future?

W: I have no idea, not a clue.

I: Do you have some guess?

W: No, I honestly don't. I mean, it's cool, but it will be some time before the quantum computers exist. It's counter-intuitive, some of these properties of quantum computers.

I: Your algorithm and other ideas on sequence alignment have also been applied to linguistics, human language development and even consumer purchasing patterns. Are you surprised by this, and do you know whether anyone has tried applying them to imaging or pattern recognition?

W: I'm not too surprised by it. I remember years earlier in Idaho trying to use the alignment algorithm to compare two different poems which clearly had a related source. I myself was trying in this direction. One of the earlier persons in this area, David Sankoff, has always had a very serious interest in linguistics, and so this connection was there all the time. For image matching, the alignment algorithm has been generalized to multi-dimensional objects — but it's not so clear how you make this work properly. Alignment has become part of pattern recognition. Motivated by the problems from biology, people in the area looked at approximate string matching, at the statistical distribution of random strings, and more and more elegant string matching algorithms appeared. This area motivated a lot of work.

I: So much of the DNA in the gene is "junk stuff" without any apparent functionality. It seems hard to believe that Nature is so "wasteful" in her designs. Could it be that there is something we don't understand about this "junk" DNA?

W: I'm sure there is a lot we don't understand about the junk DNA. In very recent years, there are all of these microRNA genes that are around and have important applications such as regulating gene expression which no one knew about a few years ago. But by the numbers I've seen, maybe 6% of the human genome is under positive selection, and we know maybe about 2% of it that is critical to the organism. There is a tremendous amount we don't know.

I: So most of the selection is not in the positive direction.

W: You are asking, "What about that other 94% or 90%?" You know, it may be like a typical mathematician's office, stacked with papers that you may never have to look at again, that you might use, that you don't throw out. I think that's the difference between the Executive Office where there is hardly clutter and the working mathematician's office.

I: Has there been any progress on this junk DNA?

W: People are all the time looking for patterns in it. Some of the answers may be in how the DNA packs into the cells, the accessibility of the DNA that initiates the copies. It is not clear.

I: Do you think that it is ever possible in the distant future to use extremely powerful computers to simulate how Nature experimented successfully with the nucleotides and other building blocks of life in producing the first primeval life form?

W: It's a question we will probably never know the answer to. They're fascinating questions. Personally, I very much like this idea that the original information molecule was RNA with DNA absent, but there's not too much known for computing the origins of life.

I: Could life have started as a kind of random process?

W: Many people believe that. In the naïve calculations, we take a protein molecule, 100 long and take 20 to the 100th power, and say that's how it happened. Or that calculation and others like it are used to argue for its impossibility. These arguments are spurious in my opinion. Just how we got the original self-reproducing molecule at the origin of life is a really fascinating question.

I: Can we do some simulation? After all, the rules of combining are known.

W: Maybe. But people also try to figure out what the environment was, of course. That is one of the key ingredients. The complexity is enormous and then there's that billion years of the early earth that you have to catch up.

I: Are there any expectations for the next great conceptual breakthrough in biology?

W: I personally don't have any predictions. I wish I did. I'm still amazed by these small RNA molecules, genes that are so important and we didn't know about them until just a few years ago. And I'm sure there will be something else like that we just haven't thought of.

I: Does it mean that the RNA is more important than the DNA?

W: Well, if first operating molecules in the cell were RNA, that would make RNA very important. One of the key pieces of evidence relates to ribosomes which are assemblies (or machines) made up of sixty some proteins and three structural RNAs. Ribosomes translate messenger RNA into proteins. Harry Noller showed that ribosomes can function without all the proteins, just with the structural RNAs. This is quite surprising and suggests to me that RNA may have been there before proteins.

I: Are there any models for the origin of life? I remember Freeman Dyson once proposed some kind of model.

W: People are always writing about that. But I don't follow it carefully.

I: What advice would you give to someone who wants to study computational biology?

W: I feel that it's important to learn as much basic chemistry, basic physics and basic statistics as the student can. The basic facts are extremely important and some depth in mathematics and biology is also required, of course.

I: Do you have any PhD students?

W: Three PhD students and one postdoc at this point. We are attracting students into this area who really come prepared. They know what they want to do and they take some serious courses... 15 years ago, people often had a degree in a different area and then converted to computational biology.

I: Are there any special programs in this area?

W: We have a computational biology PhD program within the Biology Department. The students take courses in biology, mathematics, statistics and computer science. They work hard!

Author Index

Photo Credits

We are grateful to the following for their permission to reproduce the photographs in this book:

- The School of Economics and Social Sciences, Singapore Management University for the photograph of Robert F. Engle.
- Roe Goodman for his own photograph.
- Wilfrid Kendall for his own photograph.
- The Department of Statistics, Stanford University, USA for the photograph of Charles Stein.
- The Institute for Mathematical Sciences, National University of Singapore for all the other photographs.

Disclaimer